Creative Maths Activities for Able Students

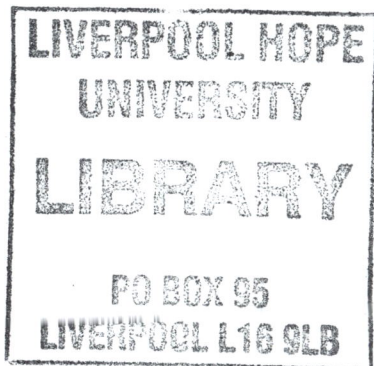

Creative Maths Activities for Able Students

Ideas for Working with Children
Aged 11 to 14

Anne Price

P·C·P

Paul Chapman
Publishing

P·CP Paul Chapman Publishing
A Sage Publications company
1 Oliver's Yard
55 City Road
London EC1Y 1SP

SAGE Publications Inc
2455 Teller Road
Thousand Oaks, California 91320

SAGE Publications India Pvt Ltd
B-42, Panchsteel Enclave
Post Box 4109
New Delhi 110 017

Library of Congress Control Number: 2005910369

A catalogue record for this book is available from the British Library

ISBN-10 1-4129-2043-4   ISBN-13 978-1-4129-2043-8
ISBN-10 1-4129-2044-2 ✓   ISBN-13 978-1-4129-2044-5 (pbk)

Typeset by Pantek Arts Ltd, Maidstone, Kent
Printed in Great Britain by The Cromwell Press, Trowbridge
Printed on paper from sustainable resources

Contents

List of figures

List of tables

About the Author

Anne studied at London University for her first degree, spending the early years of her working life as a hospital physicist in Cardiff. A part time lecturing commitment led to her move into secondary teaching.

As a mathematics teacher, Anne held a variety of additional roles, both within the pastoral system and departmental management, and undertook courses at London, Nottingham and Oxford Brookes Universities to develop her expertise and to extend her knowledge. She was fortunate during her time as Learning Support Co-ordinator to be released to undertake research for Bedfordshire School Improvement Partnership on a one day a week basis. Anne obtained Advanced Skills Teacher (AST) status in 2002.

Prior to joining Oxford Brookes, Anne worked as a Key Stage 4 consultant, developing practice in a number of secondary schools. She now works as a Senior Lecturer within CPD at Westminster Institute of Education, spending a considerable proportion of her time delivering M-level courses on Gifted and Talented education to teachers across England. Her personal research interest is in student empowerment within mathematics.

Acknowledgements

My thanks go to the many who have supported me in the production of this book.

To my tutors (now colleagues) at Oxford Brookes University who sparked off my interest in creativity.

To the mathematics department and senior team at Cedars Upper School who generously gave me the time and opportunity to trial my ideas.

To my family of able mathematicians who have made this book possible through their practical assistance, constructive criticism and ongoing encouragement.

My thanks also to the © Qualifications and Curriculum Authority for agreement to include within the Teaching Notes extracts from the *National Curriculum in Action* to be found at http://www.ncaction.org.uk/subjects/maths/progress.htm

Introduction

This is a book written for busy teachers who want to make mathematics more exciting and more demanding for the able students they teach. The photocopiable activities are suitable for use inside the classroom on a regular or occasional basis and the materials also provide useful enrichment tasks for after-school clubs or master classes.

Creativity is making a comeback on the educational stage, and for those who wish to know more, the opening chapters give an introduction to the topic and offer ideas on the education of the more able.

The target-driven culture within which you, as a teacher, are obliged to operate is not going to go away, but at least there is a growing awareness among those in power that creativity raises self-esteem, prepares pupils for life and enriches their lives. (*National Curriculum Handbook*)

The script remains the same but producers are being encouraged to be more innovative; this book is designed to facilitate that process. The creativity that is being encouraged may not be innovative when compared with that of da Vinci or Einstein, but for any individual young mathematician represents the beginnings of originality of thought.

This book has grown out of research undertaken while teaching full time in a comprehensive school. The area for research emerged in discussion with gifted and talented students themselves, who felt unable to express their creativity within the mathematics classroom, and their teachers, who felt constrained by the demands of the mathematics syllabus. These classroom needs led me to develop materials which are challenging and fulfilling for the student while not putting heavy demands on the teacher.

Terms such as 'gifted' and 'more able' have been, and will continue to be, used as interchangeable terms throughout this book as the label is not in itself important. What matters is meeting student's individual needs.

My intention is to support teachers, who feel restricted by the rigidity of the syllabi, to provide opportunities for the most able mathematicians to demonstrate what Torrance and Goff (1990) refer to as 'that special excitement' experienced when being creative. Coursework used to permit students to expand their knowledge but it is now so time limited that the bell sounds before the situation can truly be explored. To me there is a need to encourage students to maintain high levels of curiosity and playfulness as they use and extend their subject-specific expertise, while developing the communication skills relevant in our technological age.

My research findings convinced me of the need to include more opportunities for students to develop their creative skills, and I modified my own teaching accordingly. There was, and is, no doubt in my mind that, when teaching the most able, I could and should take more risks and allow them to take greater control over their learning. In a way the research gave me permission to adjust my approach as I had evidence to support the change.

Not all teachers have such a professional development opportunity and in today's classrooms the teacher of mathematics may or may not be a specialist. Diverting from the straight and narrow of any imposed curriculum requires greater risk-taking on the part of the teacher, and it is more difficult to take such risk if one lacks confidence in knowing all the possible avenues that gifted students might explore and in one's ability to follow the footsteps.

My intention is to support a move away from the view of teacher-giver and pupil-receiver model towards a teacher as facilitator approach – to me an inevitable step if mathematically gifted students are to reach their potential. Such a move does not require the teacher to have all the answers but to have strategies for finding them. The situation is then less threatening to the teacher, as there is no requirement to know all there is to know. In a democratic classroom atmosphere there is an acceptance that every individual has a part to play in reaching a solution.

The book is in two parts that may, if preferred, be read independently. Reading Part A does, however, help with understanding the nature of the tasks in Part B. Part A is an introduction to some of the background literature, considering what is meant by the term 'creativity' and how teaching attitudes might impact on its development. Part B contains photocopiable activities that may be used within the classroom or as enrichment activities primarily for students aged 11 to 14 years. While most activities have been designed for small groups of able mathematicians, it is possible to use many with the whole class or with individuals. Activities are planned so that progressively more control may be handed over to the student. How much or how little mathematical freedom is given to the student is a choice made by the teacher through considered selection of the activity.

Early chapters in Part B offer structured activities and leave the control with the teacher, but as the book progresses more and more control is handed over from the teacher to the student.

Teachers' notes give guidance on the use of the materials but the main recommendation is that examples towards the end of the book should not be used with students who have had no previous experience of freedom over the direction of their mathematical work.

The notes allow the teacher to consider in advance:

- necessary existing knowledge using the National Curriculum progress information website at www.ncaction.org.uk/subjects/maths/progress.htm
- lesson objectives (The open-ended nature of the activities means that there may well be some learning outcomes which have not been anticipated!)
- grouping
- equipment
- minimum time
- suggested outcomes.

Those teachers who use the activities frequently will notice some repetition of information, but this is unavoidable if each task is also to be accessible on a stand-alone basis.

Activities are not differentiated by year group or age because any activity could be suitable for any group. The range of ability within the 'most able' is such that you, as a teacher, are best placed to decide whether or not any particular task is appropriate.

The activities should generate their own extension activities: students will get caught up in the ideas and make their own highly creative suggestions for follow-on activities. Teaching notes include some possibilities and, in some cases, descriptions of how students have extended the material.

Students are offered the opportunity to present their findings in a variety of mediums but with a strong emphasis throughout on talk as a means of clarifying thought.

As mentioned above, the activities are designed to appeal to the able 11 to 14-year-old learner but there is really no age restriction on their use. Where the student is at mathematically, not chronologically, is the most important factor. You may decide that students significantly younger or older could benefit from a more creative mathematics experience. The bar chart in Figure 1.1 provides an indication of likely suitability across the range, but may of course be extended in both directions remembering that it is 'mathematical' age that is the most important.

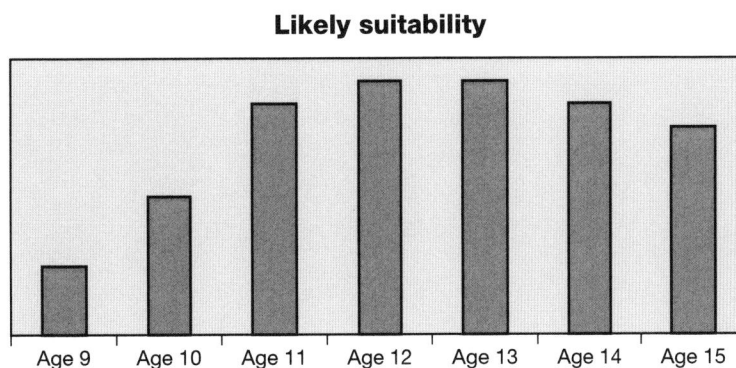

Figure 1.1 Suitability

A colleague of mine, an English teacher who would describe herself as an average mathematician, says that she now views mathematics differently having worked with me. My approach is multidisciplinary, allowing students to bring together a range of creative responses.

That is not to say that there will not be hesitancy when considering the tasks. It is likely that teachers will feel uncertain about the merits of presenting mathematics in a different light. Students, too, might need some encouragement to dip a first toe in the water but this uncertainty will be rapidly replaced with excitement in generating ideas and pride at reaching illuminating mathematical solutions. Commitment will come about naturally, as the tasks are intended to be intriguing such that the able mathematician is motivated to explore and to discover.

As Goethe is reputed to have said, 'Whatever you can do or dream you can, begin it. Boldness has genius, power and magic in it: begin it now'. The message to both teachers and students is, enjoy the challenges, the risks are worth taking, have fun and learn!

Part A

This short section of three chapters explores issues such as:

- thoughts on creativity
- the education of gifted students
- is teaching an art or a science?
- relationships, trust and risk
- assessment.

Ideas will be introduced that are relevant in connection with the notion of creativity but there is no attempt to cover topics in depth. If you wish to investigate further, references are provided.

'Gifted', 'more able' and 'able' have been used as interchangeable terms. Although I appreciate that within any school the distinctions may be relevant, my purpose is to offer suggestions to use with those students you wish to extend mathematically whatever label is attached.

Chapter 1

Perceptions of creativity

This chapter will include an introduction to:

■ Definitions of creativity

■ Thoughts on promoting creativity.

My practice is the outcome of my thought and my improved practice is an outcome of my improved understanding. (McNiff, 1993: 17)

The starting point in extending the creative dimension of classroom practice for some practitioners is to achieve a deeper understanding of what is meant by creativity. Other readers however, just as those who assemble a flat-pack item by trying it first without referring to the instruction leaflet, would prefer to move initially into Part B and return to this section later to check the details having had some practical experience of developing creativity in the classroom.

There is no one universally agreed definition of creativity but, by considering some of the writers in the field, it may be possible to replace the notion of 'knowing it when one sees it' with something more specific and perhaps more tangible. Consequently, a better appreciation of how to manufacture situations which permit 'that special excitement' which Torrance and Goff (1990) link with creativity may be acquired. Perhaps a quick scan of the flat-pack instruction leaflet before picking up the screwdriver makes the assembly less fraught with mistakes.

Choice is a fundamental theme throughout this book: decide which approach is best for you.

The word 'creativity' immediately conjures up images of artists such as Picasso and Hockney, of musicians composing and performing in a unique fashion, or perhaps of dancers moving across a stage with stunning grace. The awe factor is closely linked with creativity. Thoughts of scientific advancement, technological innovation and the like are seldom evoked by the word as, despite the frequently present awe factor, these areas assume lower status in the creativity stakes. Creativity implies 'awe', but not the reverse perhaps.

What is being judged is the product of the creative act. In the realm of the Arts it is perhaps easier for the public at large to judge the originality (and beauty) of the outcome than is the case in more technological spheres. There are few, however, who would not recognize the creativity encompassed in the designs of the da Vinci machines; machines for flight, for war and for construction. The designs are both aesthetically pleasing and fit for purpose.

During the Renaissance art and science were less distinct as areas of knowledge than is customary today, so there were fewer constraints placed on creative thought and enterprise. The bells which encourage subdivision of thinking into physics or music, mathematics or art, did not exist.

Historically, many cultures have attributed the notion of 'getting an idea' to the intervention of some higher being. This belief leads to a dampening of the need or desire to study creativity. The first systematic study of genius was by Galton in 1869, but it is the 1950s flurry of activity by Guilford and others which led to the three current research avenues of personality, cognition and development of creativity.

The belief that one must be a very special individual to have creative talent is commonly held. In his 1994 book *Creating Minds: An Anatomy of Creativity*, Howard Gardner applied his multiple intelligence approach to the selection of highly creative individuals. He suggests a link between the way in which the individual is intelligent and the nature of the creative output using Einstein to exemplify high logical-mathematical intelligence.

To understand the principles behind the activities in later sections it is helpful if you hold the opinion that every individual has the capacity to demonstrate some degree of creativity in some sphere, in this case mathematics, and that this talent can be developed given the opportunity.

This book is not seeking to develop creativity on the scale of the famous people cited by Gardner. The creativity that is being encouraged throughout this text is frequently referred to as 'little c' creativity. The ideas or products generated may not be innovative in terms of the insights available universally, but in respect of the current mathematical position of the learner represent a leap forward.

Society expects that in order to develop a sporting ability to a high standard, hours of arduous practice are needed. Why should we not approach creative talent in the same way? Olympic champions do not simply turn up and win – without personal effort and supportive trainers, medals would not be earned. Similarly the student needs to be allowed time, opportunity and coaching to foster the development of creative talent in whatever sphere.

While the expertise of a sportsman is specific to a particular event Craft et al. (2001) argue that development of creative skills provides individuals with an attitudinal set which is transferable across disciplines. Creativity is not bound by subject syllabi. Creative students grow into creative adults who will find their own unique way through situations whether within the workplace or outside it. It is not just advertising executives who need to be creative in today's world. Blue-sky days where ideas are generated free of the restrictions of practicality are common across a wide range of employment areas.

De Bono's (1985) thinking hats are used extensively within both the business environment and the classroom, with contributions to any argument made in the style of the hat 'worn':

- White hat: the facts
- Red hat: the emotions
- Black hat: the negative aspects
- Yellow hat: the positives
- Green hat: creative ideas, seeing the problem in a new light
- Blue hat: manage what is learned.

Green hats are frequently provided at governor's meetings! Creative ideas are valued and expected in numerous settings.

There is little evidence to date to give clarity to our understanding of how training for creativity might occur but many researchers feel able to list statements that encapsulate the skills necessary for the development of full creativity and those that hamper its development.

Sternberg and Lubart (1991) suggest that to foster creativity the students should be taught to work with these five resources: intelligence, knowledge, intellectual style, personality and motivation – each within an environmental context. It is but a short step from this point to a position where 'training' to enhance the skills of creativity is possible and desirable, even if the methodology remains uncertain.

If teachers take no action to improve the learning situation then there can be no consequent change. In the words of Wayne Gretskey, that famous hockey player 'You miss all of the shots you never take': we as teachers need to propel the ball towards the goal, even if initially our efforts are off target. Perseverance will lead to success.

You, as a teacher, know that the able student may engineer the situation so that he/she does not incur the penalty of additional identical problems to solve. A delayed start combined with reduced work rate allows the student to fill the allotted lesson time: introducing stalling strategies or distractions reduces the risk of repetition from the student's perspective but incurs a loss of valuable learning time.

Sternberg and Lubart (in Sternberg, 1999: 9) suggest that while to be creative requires one to be motivated, the effect is cyclical in that working creatively itself promotes a more engaged attitude to learning – a catch-22 situation. In my experience, an approach that encourages independent learning can provide a means of recapturing the interest of those who become disenchanted with instruction. This is perhaps of particular relevance when considering the education of the gifted, as for many the diet may appear dull and lacking in challenge. More of the same to fill up lesson minutes with needless repetition, unfortunately, remains a common experience.

For some students if the environment is not conducive, motivation will not be sufficiently high to result in creative working, but others will be able to cope in suboptimal conditions. Some chefs can create a banquet in a bedsit; others equally talented cannot demonstrate their skills when hampered by cramped space and poor equipment. For some the lack of a suitable environment in which to present their dishes and demonstrate their talents would be sufficient to prevent any engagement.

It is worth taking this into account when selecting activities in later chapters, as some gifted students will not engage with tasks which they know they will be unable to complete to their personal satisfaction within the time allowed.

The activities are designed for the able mathematician but there is debate about the link between intelligence and creativity, some research suggesting that up to an IQ of 120 there is a correlation between the two but that subsequently they diverge (Getzels and Jackson, 1962). Geake and Hanson (2005) working in Oxford have found evidence to suggest that people with high creative intelligence have above average working memory, enabling more links to be made and more outcomes to be considered concurrently. Fewer ideas have to be discounted from the outset in the decision-making process, allowing more 'Shall I do this or this?' comparisons.

Among the components of creativity listed by Urban (2003) are tolerance of ambiguity and task commitment. Tolerance of ambiguity is certainly key to accessing the mathematical activities described later and motivation/task commitment essential if 'solutions' are to be found. The time spent in obtaining any one 'answer' is likely to be significantly more than previously experienced by young learners.

As yet, computers have no creative capacity but many would argue that science fiction points to the future and that one day the ability will be 'learnt' by the computer. It is unlikely to happen by chance and will require the intervention of a programmer. Similarly, if left to chance, the pupil may not develop his or her true creative potential. The computer must of course be switched on to learning to learn.

It may be interesting in this context to look at the research undertaken at Bristol University into an effective lifelong learning inventory. Creativity is one of the seven dimensions used by the researchers to describe an individual's 'power to learn' and is found to dwindle over time in formal education if active steps are not taken to promote development. More details may be found at www.bris.ac.uk/education/enterprise/elli.

We are told that Leonardo da Vinci's interests were so broad that he left many paintings unfinished. The teacher's role is to balance the freedom and constraint equation so that outcomes are purposeful and of value. Too much structure and/or too much information would diminish the sought after creative impulses, too little and the pupil either pursues avenues which fail to deliver improved learning or lacks sufficient guidance to begin a creative journey. The student must be encouraged to apply both specific and general knowledge to each creative challenge.

Freeman (1991) argues that rigidity within educational systems can impact negatively on the development of creative responses. Is there a need in our knowledge-rich society to even consider adopting what Freire (1996) terms the banking method whereby knowledge is poured into empty (student) vessels by an all-knowing teacher with little if any pause in the filling-up process to work with the new information? Surely the easier it becomes to access knowledge, the more important it becomes to integrate the new with the old, creating exciting links and possibilities?

For the more able, this is consistent with working at the higher level of Bloom's taxonomy (in Maker and Nielson, 1995) employing synthesis and evaluation as discussed later. If gifted students are to have the possibility of developing into gifted adults then there must be more to their learning that the near effortless achievement of a page of red ticks. Professor John Geake states on an Oxford Brookes University Research Forum published in 2005 that students especially the most able 'should be pushed to the limits of their working memory capacity for conceptual complexity' which is obviously not the case in many mathematics classrooms.

For the most able mathematicians it is frequently possible to complete set exercises correctly with very little thought. The student's attention may wander elsewhere when there is no challenge offered and little if any genuine learning is accomplished.

Moderate challenge has been shown by Heinzen (in Sternberg 1999: x) to be most conducive to original problem-solving, while high levels of anxiety interfere with retrieval and hence inhibit the flow of ideas. Our aim as teachers is to find the position of optimum challenge. It must be kept in mind that, for those who have never had to put more than a toe outside their comfort zone, the possibility of 'failure' can prevent any risk-taking. Failure met for the first time in adolescence can be damaging to future progress if not handled appropriately. For many of the activities suggested there is no single answer, a fact which should be made clear to the students so that any fear of failure is reduced. You may observe some hesitancy and anxiety when students first experience freedom but these feelings are soon replaced with excitement at the possibilities.

There are many who would argue that the inflexibility of today's assessment-driven educational system does not permit teachers to dedicate time to cater for the particular needs of those students who in the target culture are doing very well. Talking with these students will, however, reveal how little of what they are capable of they demonstrate at school and what a loss this is to them as individuals and to society as a whole.

According to page 29 of the National Advisory Committee's report, 'All our futures: creativity, culture and education', produced in 1999, creativity may be characterized as:

- always involve thinking or behaving *imaginatively*
- overall this imaginative activity is *purposeful*; that is, it is directed to achieving an objective
- the processes must generate something *original*
- the outcome must be of *value* in relation to the objective.

The full report is available at www.dfes.gov.uk/naccce/

More recently the Qualifications and Curriculum Authority (QCA) have stated on the creativity website (www.ncaction.org.uk/creativity/whyis.htm) that 'By promoting creativity, teachers can give all pupils the opportunity to discover and pursue their particular interests and talents. We are all, or can be, creative to some degree. Creative pupils lead richer lives and, in the longer term, make a valuable contribution to society'.

To promote creativity the QCA suggest that planning should:

- set a clear purpose for pupils' work
- be clear about freedoms and constraints
- fire pupils' imagination through other learning and experiences
- give pupils opportunities to work together.

Teaching should:

- establish criteria for success
- capitalize on unexpected learning opportunities
- ask open-ended questions and encourage critical reflection
- regularly review work in progress.

Creativity should be seen in various ways such as making connections, exploring ideas and reflecting critically on ideas.

Firing pupils' imagination is central to the ideas in this book but it relies on teacher intervention in terms of asking the right questions, and requiring a reflective approach on the part of the learner. The teacher can to some extent 'give them a push and let them go' but while the choice of route is up to the student, the teacher will have a journey's end in mind. Student reflection en route, prompted if necessary by teacher questioning, avoids too many fruitless turns.

?	**To think about:**

What would your students say if you asked them how and when they are able to be creative within school?

🖱	**If you would like to know more:**

www.edwdebono.com
www.ncaction.org.uk/creativity

Treffinger, D. (ed.) (2004) *Creativity and Giftedness (Essential Readings in Gifted Education)*. Thousand Oaks, CA: Sage Publications.

Fisher, R. and Williams, M. (eds) (2004) *Unlocking Creativity: Teaching Across the Curriculum*. London: David Fulton.

Chapter 2

The education of gifted students

This chapter looks at:

- Models which can be applied to the education of the gifted
- The acceleration enrichment debate.

The situation for gifted mathematicians may be improving but there is still a significant shift in attitude necessary, within the school environment as opposed to summer schools and other extra curricula activities, if these pupils are to be allowed to reach their full potential.

David George (1997: 65) states: 'One of the goals of gifted education is to develop creative and imaginative thinking as well as problem solving ... The learning experiences should integrate cognitive skills, affective skills, intuition and talents in a specific area.'

This section will touch lightly on some of these issues looking at both cognitive and affective taxonomies, but with no attempt to provide a comprehensive or detailed account of the range of possible approaches. Of those chosen for inclusion several have application across all age ranges and all abilities, but our specific focus will be on the more able.

Teacher familiarity with Bloom's cognitive taxonomy (Figure 2.1) is increasing. It has stood the test of time as a useful tool and features in many government publications such as those encouraging appropriate questioning. You may already be regularly planning higher-order questions requiring the application of skills from the top of the pyramid to promote the learning of the more able.

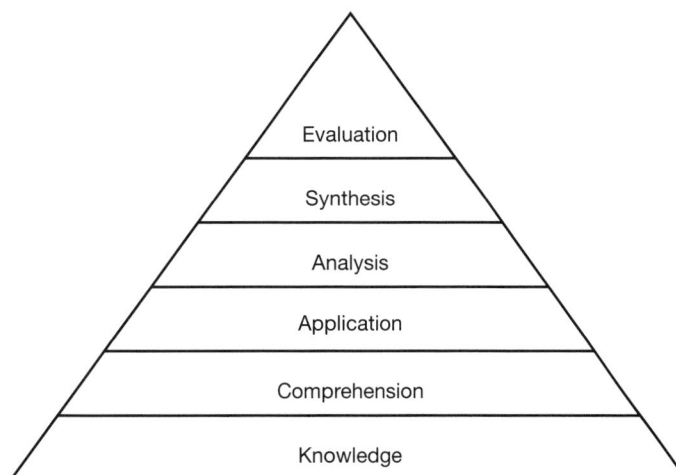

Figure 2.1 Bloom's cognitive taxonomy
From Benjamin S. Bloom et al., *Taxonomy of Educational Objectives*. Published by Allyn and Bacon, Boston, MA. Copyright © 1984 by Pearson Education. Used with permission of the publisher.

A firm basis to the pyramid is established through knowledge and understanding but the more able mathematician should be engaged in tasks which develop the skills of analysis and beyond. Synthesis and evaluation cannot be replaced with additional repetition of tasks from the lower levels.

It is these skills at the top of the pyramid that the suggestions in later chapters seek to develop. Many activities will result in increased mathematical knowledge; all demand increased understanding and application in new (and somewhat unusual) situations, but the most significant observable changes should be in higher-order thinking.

Beside the cognitive sits the affective taxonomy devised by Krathwohl and Bloom in the 1950s (in Maker and Nielson, 1995). This taxonomy looks at the emotional and attitudinal side of learning, and begins with the stages of receptiveness, progressing to a point where new behaviour is firmly embedded in the student's value system.

- Receiving: willingness to hear
- Responding: participating; questioning
- Valuing: attaching worth
- Organizing: prioritizing values
- Internalizing values: having a value system that controls behaviour

The tasks in this book are designed to be engaging so the first steps are readily taken and progression through the stages rapid.

A teacher told me recently of a 9-year-old pupil who had moved through all five stages from receiving to internalizing, in the course of a single lesson. The teacher had used synectics, an approach used to encourage creative ideas, to support a class writing session. The pupil scored very well in this area of his English SATs and explained without prompting how he had employed synectics to help him to write an exciting story. Details of this approach can be found in Joyce et al.'s (2002) *Models of Learning-Tools for Teaching*.

My intention is that through active involvement with the tasks, the students will be able to internalize problem-solving approaches in mathematics and be able to draw on these as necessary.

Maslow's hierarchy of needs (Figure 2.2) may be useful to consider. The suggestion is that individuals require their needs to be met at the lower levels, physiological and emotional, in order to enable them to move through belonging and self-esteem to self-actualization.

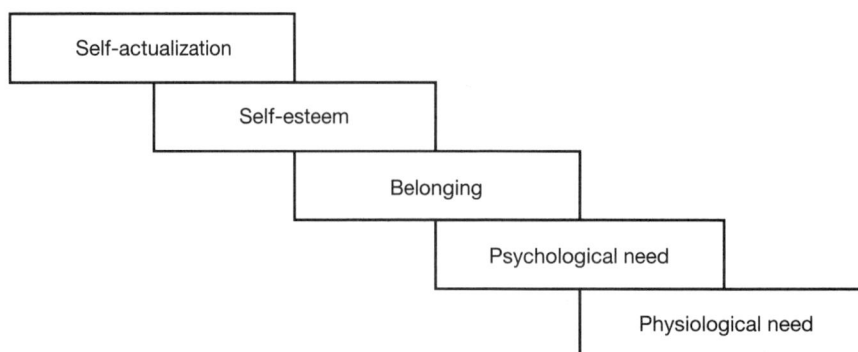

Figure 2.2 Maslow's hierarchy
© Maslow, A.H. et al., *Toward a Psychology of Being* (1999).
Reproduced with permission of John Wiley & Sons, Inc.

Self-actualization is often described as becoming everything that one is capable of becoming: self-fulfilment is sometimes used as an alternative term. Creativity is seen by many as integral to the process but not everyone agrees with Maslow as it is possible to find many examples of starving artists who fail to have their basic needs met and yet achieve fulfilment.

However, for all school students we have a duty to provide a classroom climate which guarantees that basic physiological and psychological needs are met and to compensate as far as is possible for known detrimental external factors.

As a classroom teacher, you will not have any control over what happens outside school. It is possible that students cannot perform at their best because they arrive at school hungry or tired; this is as true for able students as it is for those who find mathematics difficult, and may hamper the school work of the gifted so significantly that their ability goes unnoticed.

The reptilian brain shuts down our intellectual capacities if threats to our safety are perceived. These threats may not appear life threatening to an onlooker but need to be understood from the perspective of the individual. A student who is worried about the quality of the presentation of his/her homework to be handed in at the end of the lesson may be unable to learn, however able.

For some able students the sense of belonging is absent or diminished relative to the camaraderie experienced by classmates. Co-operative working required by several of the activities can alleviate this difficulty if sensitively handled. Change may be slow but over time it is to be hoped that group members appreciate each other's strengths and support each other's weaknesses.

Perhaps this quote taken from a letter to the Pope in Rome illustrates Maslow's argument: 'The bearer of these presents is Michelangelo the sculptor. His nature is such that he requires to be drawn out by kindness and encouragement but if love be shown him and he be treated really well, he will accomplish things which will make the whole world wonder.' The Sistine Chapel provides our evidence.

Freeman (1991) is concerned that the emotional needs of the students should be met such that they are given the confidence to express themselves in their own way. This may be challenging within the classroom environment but, with no wrong answer possible and a range of possible outcomes, the activities attempt to make life a little simpler for the busy teacher.

We now move on to consider structures through which we may provide for the gifted mathematician; first, enrichment.

Renzulli (1977 in Maker and Nielson, 1995) is well known for his theories of intelligence and models of provision. This is the only approach to be considered in this section that is specific to the education of the gifted.

His original triad model, developed in 1977 to provide a differentiated approach for the gifted, added as enrichment a layer of investigative study of real problems. The student might, for example, adopt the role of an architect offering designs for a genuine project. The outcomes are given the same level of consideration as those produced by the full-time professionals, with highly motivating effect.

In 1985 Renzulli and Reis modified the earlier version to create the schoolwide enrichment model, which as the name suggests draws a greater proportion of staff and students into the Type 1 and 2 (see Figure 2.3) activities, lessening the risk of the accusation of elitism.

In Type 3, students are encouraged to be independent and the work can diversify into many directions, some of which may be unfamiliar to the teacher who acts as a facilitator rather than a repository of knowledge. The demands on the classroom teacher are therefore different, some would argue greater, than those required by less 'free' approaches.

Type 1	Type 2	Type 3
Introduction to new and exciting topics not in the normal curriculum	Methods designed to develop higher-order thinking	Students become the investigators, formulating the problem and the methods for solution. The teacher's role is to be a manager of the learning process

Figure 2.3 Renzulli's triad
Used by permission of Sage Publications, Inc.

All activities in later chapters would offer at least a Type 2 experience with significant demands on higher-order thinking. The activities in Chapters 6 and 7 include a Type 3 aspect, with the students increasingly setting their own problems in response to the stimuli.

It may be of concern to you, as an already overburdened teacher, that you might be required to have all the answers to all the possibilities but, according to Eyre, 'encouragement of innovation and risk-taking often leads to high quality work and does not require additional teacher time' (Eyre, 1997: 93). Those familiar with coursework in mathematics will appreciate that while some understanding of where the task might lead is useful, it is not essential to have a detailed solution to every case. Data-handling coursework has such a range of possibilities that teachers are unable to explore all avenues. Within this coursework situation many teachers are comfortable to loosen the reins but while the opportunity for freedom exists many schools remain hesitant to let go completely because of possible implications for results.

It is just a small step from the coursework situation to accept that as a teacher working with the most able mathematicians, you will have an understanding of the processes but not an answer book or a file of solutions.

Decisions as to whether the enrichment activity takes place within or outside the classroom, within or outside the school day, rest with you as the teacher. If you are teaching in a small school, then linking with others in similar situations may provide a dynamic and viable group of young mathematicians.

Is acceleration an alternative? When on holiday we have our list of 'musts', and a list of 'maybes'; we plan our excursions to get the best possible experience we can. A whistle-stop tour suits some; time spent absorbing the cultural diversity suits others. What approach best suits our able students as they journey through their education? In what ways is it beneficial to arrive early at the final destination?

There have been many debates regarding this issue. A seminar held by the United Kingdom Mathematics Foundation (UKMF) in May 2000 to discuss the acceleration/enrichment issue, considered these questions set by the Gifted and Talented team from the Department for Education and Employment (DfEE):

■ What do we mean by 'acceleration' and 'enrichment'?
■ Can one base a national policy on 'acceleration' and 'enrichment' as parallel strategies?

- Is acceleration wrong in principle in all cases – if so why?
- If 'acceleration' is sometimes acceptable, what are the circumstances?
- What other conditions are needed for acceleration to be effective?

You will have your own answers. The 23 representatives from various spheres in education, with Tony Gardiner of Birmingham University as secretary, came to clear conclusions and appear to have found agreement relatively easy to reach.

'Acceleration' they defined as 'a strategy (of feeding to students) standard curriculum work months or years ahead of their peers, thereby

(i) putting their learning (permanently) out of phase with that of their peer group, and
(ii) creating a potential vacuum … when the available standard work is completed'
(UKMF, 2000: 11).

The phrasing of the definition in itself coveys negativity and while the advice from the seminar was 'if in doubt, don't (ibid.: 16), there was an acknowledgement that there might be an occasional local need.

I have personally taught small numbers of 14- and 15-year-old accelerated students who have undertaken AS/A level modules in Years 10 and 11 with mixed results. The difficulties arose not with the mathematics but with issues of transition which I regrettably failed to acknowledge.

When students move from Year 11 into 12, from Key Stage 4 to 5, the move is planned for. Frequently there are trial lessons for students so that they may get a flavour of what is to come and ask questions about the course. Moving from Year 10 into a Year 12 class with no preparation for the change was for some extremely difficult.

It is not simply a matter of teaching the student harder mathematics. In my case the move involved working with a sixth-form group in a classroom atmosphere very different from that operating in Year 10. There was a significant change in the level of independent working required and a loss of peer interaction and support. For one individual the pleasure in learning more advanced mathematics more than compensated; for another it did not.

The modular system in Britain has to some extent reduced the concern over what next, as there is almost always another module and another area on mathematics which could be studied at AS or A2 level. The student, however, remains permanently out of phase with their own year group.

If instead the accelerated programme takes place within the student's current classroom, the teacher input may be rushed as he/she alternates between syllabi, there is reduced opportunity for reflective discussion and a general dumbing down of the vitality of the subject.

Ultimately the decision as to whether or not acceleration is appropriate has to be located in the context of the individual student, to determine whether for a particular child in a particular school it is the best educative approach. Consideration of both the cognitive and affective status of any individual determines the appropriateness of the form of provision.

The 14 to 19 agenda and the recent focus on personalized learning may have an impact on how frequently such decision-making is undertaken but few would disagree with the attendees of the UKMF seminar who felt that the cognitive consequences of any special programme should include:

- stronger foundations
- deeper understanding
- a greater willingness to reflect on the connections.

The cross-curricular activities in this book seek to include and develop all three cognitive consequences; in particular, the need to make connections is emphasized as students are encouraged to apply mathematics to unfamiliar and perhaps unlikely situations.

The emphasis is on an enrichment approach but, as the student is able to exercise choice, to achieve a satisfying, creative solution, his/her learning may be extended beyond the boundaries of the curriculum stage.

? **To think about:**

Would it be possible to talk to your able students about their experience of mathematics?

If you would like to know more:

National Academy of Gifted and Talented Youth at www.nagty.ac.uk
National Association for Able Children in Education at www.nace.co.uk
National Association for Gifted Children at www.nagcbritain.org.uk
Oxford Brookes University at www.brookes.ac.uk/go/cpdgifted

Eyre, D. (1997) *Able Children in Ordinary Schools*. London: David Fulton.

Maker, C.J. and Neilson, A.B (1995) *Teaching Models in Education of the Gifted*. Austin, Texas: Pro-Ed.

It is interesting to note the extent of legislation and government activity in the area of gifted and talented education. Some of this activity is set out below.

Table 2.1   Government directions

1997	The Government publish *Excellence in Schools*
1998	Ofsted survey of provision, undertaken by Professor Freeman Estelle Morris announces the new advisory group on gifted and talented
1999	Excellence in Cities set up House of Commons Select Committee publish *Highly Able Children*
2000	Maths Year 2000 The enrichment approach is better supported than previously, for example: NRICH www.nrich.maths.org.uk/public/index.php MOTIVATE motivate.maths.org Mathematically Promising www.meikleriggs.org.uk
2001	The Green Paper, *Schools: Building on Success* including among other proposals: Early Key Stage 3 entry allowed – pilots to be set up Setting up of express sets will be encouraged to allow early GCSE entry Creation of National Centre for Gifted and Talented youth
2003	Creativity: Find it Promote it Website launch: www.ncaction.org.uk/creativity Excellence and Enjoyment published www.standards.dfes.gov.uk/primary/publications/literacy/63553
2004	Launch of London Gifted and Talented www.londongt.org/homepage/index.php

Chapter 3

The role of the teacher

This chapter will focus on the role of the teacher:

- Is teaching an art or a science?
- Trust and risk
- Assessment.

Hopkins and Harris (2000) remind us that once the classroom door is shut, creating the optimum conditions for learning becomes the responsibility of the teacher. What this means in practice is that you, as a teacher, make decisions about how learning is best promoted. This may include turning unexpected events into gainful learning experiences: questions asked by pupils may change the direction of the lesson.

Using the term 'artistry' to describe this aspect of teacher behaviour, Hopkins and Harris draw on the words of Lou Rubin (1985) from which I have selected these phrases.

Students are caught up in the learning
Excitement abounds
An unmistakable feeling of well-being prevails
Control that is born out of perceptions, intuition and creative impulse
(Rubin, quoted in Hopkins and Harris, 2000: 8).

Would you like to be the teacher in this classroom? For me this conjures up the ideal. By contrast these words from Laurie Angus (1993) draw a very different picture:

... educational practice is conceived of in a particularly mechanical way ... it is the bit between 'input' and 'outputs'. It is seen largely as a set of techniques ... Practice is imposed rather than negotiated or asserted: it is a set of techniques to be employed by teacher technicians on malleable pupils. (ibid.: 337)

This is an image of schools as education factories with production lines set out to construct identical products. Quality control inspectors see differences as defects, but surely we are not seeking to create future citizens with thought processes indistinguishable from each other. Students may be malleable as a consequence of the environment in which they are placed, although in practice many (some who are gifted) rebel against the regime. Do we desire a robotic workforce or blue-sky thinkers? How can individuals genuinely participate in a democracy if they have no experience of thinking for themselves?

The mechanistic attitude towards teaching has arisen partly because the opinion held by many is that there exists a larger number of individuals capable of acquiring 'the applied science of a teaching methodology' as compared with the excellent few who inherit the art (Muijs and Reynolds, 2001: viii). In my opinion, we start out in teaching with a mechanistic approach but as our experience grows we develop the artistic side more and more.

The National Strategies emphasize equality, and I am sure that we would all agree with the desire to provide every child with good quality education. Equality does not mean identical learning experiences and I would encourage you to bring your artistry into play and allow excitement to flourish.

In our data-rich society information is readily available. We need to focus on developing research skills which allow students to find out basic facts they may need to know before moving on in their learning. With the teacher they then work with the facts to turn them into knowledge. Artistry puts the skills of teacher and student to use in ways which bring the classroom alive and it promotes the excitement in learning experienced by a young child taking his/her first steps. The joy on a toddler's face is seen far less often if ever on an 11-year-old but excitement is contagious and can infect whole classes: magic is made as teacher and taught journey together.

Should we then as teachers strive to retrieve our artistry and 'like jazz musicians react to circumstances on the spur of the moment' (Humphreys and Hyland, 2002: 10)? As must be obvious, I believe that teaching is much, much more than a set of competencies: to me artistry and the associated creativity are integral to the scene.

We do not hold all the answers but in a knowledge rich world the 'answer is no longer the key to the process. The key is the question' (Dalin and Rust, 1996: 145, cited in Day, 2001: 491). The activities in this book encourage questioning and bring together many areas of the curriculum not dissimilar from 'topic work' undertaken pre-strategies. Primary teachers tell me that it is coming back and those who have been teaching for many years are heard to say 'What goes around comes around'.

The role of teacher is creeping towards the role of facilitator, but the steps are slow when society continues to value demonstrable teacher knowledge above teaching artistry. There is, of course, a clear need for teacher knowledge and competence: to ascend Bloom's hierarchy the foundations must be firm. Jazz musicians acquire the skills and understanding associated with playing their instruments before they begin to improvise, so too must the mathematics teacher acquire a sound understanding of the more controlled situation before 'letting go': every activity should result in meaningful mathematical learning for able students.

Jazz musicians do benefit from being encouraged by their audience to be creative with their knowledge rather than to merely repeat a series of set moves. I can think of one occasion when a class clapped my 'performance', but only one in a long teaching career! The after-school class was a group of Year 9 able mathematicians I had never met before and, although somewhat surprised, I was delighted by their response.

We have to look for encouragement of a less overt nature; the parent who tells us of the ongoing work at home, of the enjoyment experienced, the student who just wants to keep going regardless of the bell.

Teachers tell me of their anxiety to work within the national strategies which they frequently perceive as restrictive. They need support from management if they are to weave the learning objectives into a more integrated learning experience. Concerns regarding the attitude of Ofsted whether justified or otherwise have kept schools in the one-way system but I have visited one primary school which teaches the curriculum through creativity with notable success. The National College of School Leadership (NCSL) have produced a document 'Developing creativity for learning in the primary school: a practical guide for school leaders' which may be found on www.ncsl. org.uk/media/ F7B/93/randd-creativity-for-learning.pdf.

Writing 40 years ago, Bruner (in Maker and Nielson, 1995) considered that we had less interest in encouraging intuitive thought processes and were increasingly valuing logical and systematic thought. Over recent years, the ability to retain and repeat facts, to become dispassionate and detached, has been emphasized at the expense of learning by doing.

Computers retain and repeat facts devoid of passion: people bring emotion and beliefs to those facts. In the twenty-first century the medical profession are increasingly turning to computers to supply facts as the body of knowledge becomes larger and more complex, but the machine does not replace the medic when it comes to deciding how to interpret the facts presented.

National strategies do have benefits to offer, especially equality of curriculum content, however, the nature of the exposure to the content does (and in my opinion should) differ significantly. Ideally, every child's individual learning needs should be met without disadvantaging the student in an assessment-driven culture.

For me there is a balance to be struck. The scales must not be weighted too far towards recall, hindering the development of creativity and increasing the risk of demotivating those with the potential to be able mathematicians. For the most able the simplicity of a question requiring only recall, may lead to confusion, as the instinctive response appears too simple and hence incorrect.

Gardner feels that one of the factors which predispose people to be creative is early exposure to others who are comfortable with taking chances. For some students this exposure may be available to them within the home; for others it may not. If parents are unable to provide a creative role model then it becomes even more important that teachers do so.

Schools demonstrate value in the 'the right answer' through teacher response and test scores. Able pupils from an early age are expected to achieve the perfect score with little time built in for the what-ifs; many are discouraged from questioning 'why' in maths lessons. While there is now more encouragement to value alternative approaches to reaching the right answer, there is still the sense that teacher knows best. 'Never mind that, just do it my way and you will get it right!'

To encourage creativity the school must have an obligation to suggest that learning is as much a process as an end result. We need to present learning as a journey that does not always result in an immediate correct response and does not always have an unbroken straight line between start and finish. Gifted students may have had no previous experience of any wrong turns or dead ends so at an early stage of their learning will need reassurance that seemingly blind alleys may ultimately lead to more exciting destinations.

We all know of gifted students who are devastated by failing their driving test, the first examination situation in which they have not succeeded. Nothing in their school experience has prepared them for the feelings they experience.

Buescher and Highman (1990: 2) point out that 'While risk taking has been used to characterize younger gifted and talented children, it ironically decreases with age, so that the bright adolescent is much less likely to take chances than others'. Our aim therefore is to provide a classroom environment where it is OK to fail and where stepping out of one's comfort zone is normal practice. By providing 'safer' risk-taking, green hat thinking may be a more common event.

As de Bono says, 'If creativity is no longer a risk then non-risk takers may decide to be creative' (de Bono, 1982: 55). However, while 'the fear of being wrong and the fear of making mistakes prevents the risk-taking of creativity ... Releasing a brake on a car does not automatically make you a skilled driver' (de Bono, 1992: iv).

In fact, releasing the brake on a vehicle perched on a hill top with the engine off may have a less than desirable effect! Ensuring that the car is safely placed and with the engine well serviced, effective brakes and adequate fuel are all necessary before speeding forward into new and exciting lands. There will be hills to climb, so the engine capacity must be sufficient and, of course, if it is dark the teacher needs to show the pupil where the head-light switches are.

De Bono does agree, however, that a mild level of creativity is encouraged simply by removing constraints, and the ideas in later chapters are designed to provide increasing freedom so that the teacher instructor/facilitator can step back a little further as the student becomes more efficient at driving his/her own car.

During the 1950s the USA promoted research into scientific creativity. Calvin Taylor (in Maker and Nielson, 1995), a psychologist developed his multiple talent approach suggesting that about a third of students would exhibit giftedness in one major talent area. The talents teachers are challenged to develop are:

- creative
- productive thinking
- forecasting
- decision-making
- planning
- communication.

The model provides a positive way of viewing pupils and appeals to many teachers because of its inclusivity. Many of the teachers I work with adopt it as an assessment tool using Taylor's totem pole approach (Figure 3.1).

Creative	Productive thinking	Forecasting	Decision-making	Planning	Communication
Fred	Jill	Said	Jill	Fred	Susan
Susan	Fred	Fred	Said	Said	Jill
Jill	Susan	Susan	Fred	Susan	Said
Said	Said	Jill	Susan	Jill	Fred

Figure 3.1 Taylor's totem pole
From *Creativity and Giftedness*, (2004) Treffinger, D. (ed). Used by permission of Sage Publications, Inc.

Students are placed in rank order under each heading so that strengths and relative weaknesses are clearly identified and can be used to plan for individuals by, for example, allocating different roles in group work. It would be possible to devise a similar totem pole for specific mathematical skills taking, for example, Level 7 or 8 National Curriculum descriptors as the headings.

The model emphasizes transferable skills, rather than content, leading, hopefully, to more rounded adults. Talents are not mutually exclusive, so learning activities tend to develop a cross-section. As a consequence, in parallel with other provision, Taylor's approach effectively enhances both achievement and creativity.

To be successful in promoting creativity you will need to use your own well-developed thinking and creative skills and to consider students as thinkers rather than knowledge absorbers.

The relationship between yourself and your students may also undergo change, but this will depend on your overall teaching style. Relationships are always complex and the first few steps difficult to take, but unless risks are taken there can be no movement from the status quo. Steps may initially be small but grow as trust develops on both sides.

The first occasion on which you, as a teacher, openly acknowledge not knowing the answer is likely to cause some initial unease. However, as there are no exact answers to the activities, as you work with the group or individual to overcome a difficulty any uncertainty will disappear. The tasks in this book are designed so that very soon any status issues are forgotten and all involved are caught up in the excitement of mathematics.

Working alongside the student, modelling more adult thought processes, enables the teacher to connect with the student's thinking in a way which is impossible as an observer. Mutual respect will grow out of the joint working and a deep interest in the subject matter.

The issue remaining is how to assess the learning. Assessment for learning involves gathering and interpreting evidence about students' learning and then acting upon the information gained in terms of moving the learning on. The teacher facilitator is in an ideal position to utilize these suggested approaches from Paul Black:

- providing feedback to pupils
- actively involving pupils in their own learning
- adjusting teaching to take account of the results of assessment
- recognizing the profound influence of assessment on motivation and self-esteem of pupils
- considering the need for pupils to be able to assess themselves and to understand how to improve.
(Assessment Reform Group, 1999: 4, 5)

Assessment can then be ongoing and continuous with no need to take in the books to add red ticks, marks or comments. It has been found by Paul Black and his co-researchers that if a mark and a comment are provided, the comment, however thoughtfully constructed to move the learning on, is ignored (Black and Wiliam, 2002).

Written products have historically been judged on neatness as much as quality in mathematical terms, and this may have disadvantaged some of your students who find the process of transmitting their thoughts into words on paper difficult. Not all able mathematicians appreciate the need to write explanations in full, believing a few scribbled symbols will suffice. Many individuals will benefit from being able to present their mathematics in a range of media.

The activities in later chapters will generate outcomes in a variety of forms which will perhaps lead to concerns regarding assessment. As is well known, it is customary to measure achievement in mathematics through written assignments. Most often these are routine exercises with which the more able mathematician has little difficulty. Occasionally, as with coursework, a more open-ended task is set but time constraints set by examination boards essentially guarantee that the majority of products are predictable from the teacher's perspective. An additional requirement usually stated is that the work is presented in such a way that any reader is capable of accessing the findings, so a significant proportion of the submission is given over to drawing diagrams and writing lengthy explanations, which can be very tedious but necessary in examination board terms.

While clarity of thought is to be encouraged, and the ability to explain ideas to others essential, the writing-up process does reduce the time spent on mathematical activity. I think that this reflective process is important in itself but that the time devoted to it is frequently disproportionate relative to the whole task, with much attention being paid to neatness of presentation. The number of pages consigned to the bin and rewritten is indicative of the importance attached to this aspect by the student.

For examination purposes this may be unavoidable and the reward for the student following the rules is high grades, but for many able students jumping through the hoops to optimize their marks is frustrating and takes away from them the time which might be productively spent deepening their understanding of the mathematics.

Jensen suggests that 'You either get intrinsically motivated creative thinking or extrinsically motivated repetitive rote behaviour' (Jensen, 2000: 266). Intrinsic motivation achieved through immersion in a problem is more likely to be possible outside the examination system. By encouraging a more creative approach at a younger age, the student should be able to achieve more when the assessed situation arises.

Removing constraints on the way in which learning is demonstrated can be highly motivating and encourage a more creative approach to many tasks. Many researchers agree (Nickerson in Sternberg, 1999: 417) that freedom to choose is a prime motivator leading to enhanced creativity.

Just a simple change, such as giving students the chance to explain their thinking to a teacher or a peer as an alternative to the medium of writing, could release useful time for mathematical activity as well as providing an opportunity for reflection.

As will be discussed in greater detail in later chapters, students given such freedom choose to present their work in a range of ways. Products such as raps, board games, diaries and photographic evidence can be expected alongside the more traditional experimental work, frequently presented as websites using highly developed information and communication technology (ICT) skills. Suggested outcomes are provided for each activity but the greater the freedom you offer, the more inventive will be the response.

The difficulty which remains is how to measure the relative worth of the products. The decision has to be made about whether there is a need for measurement as such, as it is the learning that has occurred through the process that is of importance as much as products themselves. The learning can be quantified through more traditional methods if a score is necessary.

That is not to say that the students should not receive appropriate feedback on their work but that a comment-only approach is more likely to promote creativity. 'Creativity is rarely measured in relation to a reward system – in fact the two are usually at far ends of the same scale' (Jensen, 2000: 266).

Suggestions for outcomes for activities provided in Chapters 4–8 include debates, creating adverts and producing a script for a television programme all of which would be difficult for a mathematics teacher to mark in a conventional way but clearly demonstrate the level of learning and depth of understanding.

Peer assessment is an approach which can move forward the learning of both assessed and assessor. Students need to be inducted into the approach and taught to focus on the use of higher-order skills. It may then be used as either a formative or summative tool and support moves to greater independence.

As there are no answers the decision about the level of learning ultimately rests with you as teacher.

? To think about:

How does your teaching encourage risk taking in learning?

If you would like to know more:

Good Assessment in Secondary Schools at
www.ofsted.gov.uk/publications/docs/3205.pdf

Excellence and Enjoyment at www.standards.dfes.gov.uk/primary/publications/
literacy/63553

Black, P., Harrison, C., Lee, C., Marshall, B. and Wiliam, D. (2002) *Working Inside the Black Box: Assessment for Learning in the Classroom.* London: NFER-Nelson.

Part B

This is the section of the book that provides activities which you may photocopy and use with students. Teaching notes provide guidance on:

- prior knowledge
- lesson objectives
- timing
- equipment
- suggested outcome.

In most cases questions to start the students thinking about the mathematical issues raised are suggested, but this book is unlike a standard mathematical text, so nothing is set in stone. Within each task there is considerable flexibility.

Chapter 4

This chapter has suggestions on the use of a range of teaching strategies you will find useful with all students but are much enjoyed by able mathematicians. These are:

- co-operative group work
- inductive teaching
- Direct Attention Thinking Tools (DATT)
- use of ICT.

The tasks are structured but with the opportunity for the most able to demonstrate creative approaches.

Chapters 5–7

These chapters provide tasks which become increasingly more student led, chapter by chapter. You are advised not to use tasks from Chapter 7 if your students have not had any previous experience of this sort of mathematics, as they may lack the confidence to make the decisions required of them. Examples in Chapter 5 are more structured and so reduce any feelings of uncertainty which may occur at an early stage.

Chapter 8

This has additional suggestions for master-class activities. The difference between these activities and others is more about the time needed to make some inroads into the task than the difficulty.

Chapter 4

Teaching strategies with practical suggestions

This chapter will provide information and practical examples on:

- Co-operative group work
  - Igloo

- Inductive teaching
  - Letters
  - Graphwork

- Direct Attention Thinking Tools (DATT)
  - A Porsche or not
  - The taxman sees all

- Use of ICT
  - Tug of war
  - Spheres
  - Polygons.

Co-operative group work

It would detract significantly from the purpose of developing creativity if you decided to deliver the activities in latter chapters from the front using a highly directed, step-by-step, teaching approach. That is not to say that there may be times when you will need to discuss particular points with everyone involved in order to move the learning on, but too much direction may discourage the more divergent thinker and hinder his/her creative process.

Many of the activities could be undertaken using a co-operative group-work approach to ease the transition from teacher led to student led. When researchers such as Slavin (1995) compared student achievement in small-group settings with traditional whole-class instruction they found that there were significant learning gains in the grouped situation. Thinking back to the affective issues raised in Chapter 2, you may wish to encourage group-working for social-emotional reasons as well as mathematical fulfilment. You may feel that by allocating roles within groups you will be able to cultivate less well developed talents.

Webb (1991) found that achievement was raised through the process of explaining thoughts and ideas to other group members but that receiving no feedback for those ideas had negative impact. Think–pair–share is now regularly used in British classrooms and co-operative group work is essentially an extension of this process. The skills of working as a member of a group have to be learnt, and for some able students the process of sharing ideas may be more challenging than the mathematics itself. Organizational and communication skills are developed alongside mathematical skills providing an important basis to more creative activity.

The construction of the groups needs consideration if all members are to be given equal learning opportunities. Equality of learning opportunity does not necessarily imply that each pupil engages in identical tasks. It may be useful to use differently constructed groups, dependent upon the type of activity/task (these terms are used interchangeably) and the possible learning gains. As you will notice in the example, the Igloo, each group has a roughly equivalent subtask to undertake but this need not be so. It is possible to design activities such that the subtasks can appeal to students with differing talents, each then bringing particular expertise to the main task.

Although the approach might vary, the method encourages the development of both social and mathematical skills. Each group member has individual responsibility and responsibility to the group as a whole. The jigsaw approach which is used in the example, the Igloo, requires that the group ensures that every member has the understanding necessary to work as an expert on one particular aspect, for example finding volume of a sphere.

The diagrams demonstrate how the groupings work in the classroom.

Group A

Expert 1 Subtask A	Expert 2 Subtask A
Expert 3 Subtask A	Expert 4 Subtask A

Group B

Expert 1 Subtask B	Expert 2 Subtask B
Expert 3 Subtask B	Expert 4 Subtask B

Group C

Expert 1 Subtask C	Expert 2 Subtask C
Expert 3 Subtask C	Expert 4 Subtask C

Group D

Expert 1 Subtask D	Expert 2 Subtask D
Expert 3 Subtask D	Expert 4 Subtask D

The number of groups and the number of students within each group obviously will vary according to your own class situation and the total number of pupils involved. It may be that you have just two subgroups, eight students in total, or even just two pairs, but the principle remains the same. Four is the minimum number for a jigsaw activity.

The subtasks in the example are structured but retain the opportunity for the most able to demonstrate creative approaches. The students are regrouped after the subtasks have been completed to form new groups of experts to work on the main task. Sometimes, dependent upon numbers, there will be more than one expert on a particular subtask in each main group but, as illustrated, the group now has at least one expert on each possible aspect of the main task:

| Expert 1 | Expert 1 | Expert 1 | Expert 1 |
| Subtask A | Subtask B | Subtask C | Subtask D |

This is designed as a jigsaw activity with the preparatory work undertaken first.

Ideally the students need to be divided into four groups of four or more students and each group has to acquire expertise in one particular skill by undertaking one subtask. The worksheet gives four suggestions for subtasks but you could choose to use any similar situations which arise naturally within the classroom. The sphere is likely to be the shape that the students are least familiar with.

You may decide to give measurements but it is good practice for the students to estimate appropriate dimensions.

Igloo

T Teaching Notes

Prior knowledge

Pupils need to be able to

- understand and use appropriate formulae for finding circumferences and areas of circles, areas of plane rectilinear figures and volumes of cuboids when solving problems (Level 6)

- calculate lengths, areas and volumes in plane shapes and right prisms (Level 7)

- solve numerical problems involving multiplication and division with numbers of any size, using a calculator efficiently and appropriately (Level 7)

Lesson objectives

To encourage pupils to

- calculate lengths of circular arcs and areas of sectors, and calculate the surface area of cylinders and volumes of cones and spheres (exceptional performance)

- use mathematical language and symbols effectively in presenting a convincing reasoned argument including mathematical justifications (exceptional performance)

Suggested outcome

- A poster from each group used to display their findings and as a visual aid when describing their thought processes to others

Groupings

Small groups of 4 with similar strengths

Timing

Minimum 2 hours

Equipment

Access to textbooks or the Internet

It is the responsibility of the subgroup to make sure that each member is confident in the technique and capable of explaining it to a fellow student who has not been a member of that particular expert group. Your judgement is needed to decide how long should be allocated to the preparatory work as it depends on how much prior knowledge is held collectively by the group. The time limit for the preparatory work needs to be stated at the outset.

The subgroups are then reformed so that there is at least one expert on each mathematical skill in each of the new groups. Allow time for the experts to explain to the other members how they undertook their task and then provide the igloo for the group to work on. At this stage it may be useful to engage in a brainstorming activity to discuss the shapes of which the igloo is composed.

Suggestions for questions which may be needed by some pupils at the initial stage:

- What shapes can you see in this drawing?
- What does it look like where the shapes join?
- How tall must the entrance be to let an Eskimo crawl in?

The shape can be thought of as made up of two cylindrical sections and a hemisphere. As the task is designed for able students, no measurements are given and some aspects of the shape are not defined. Apart from likely dimensions, students will need to decide the overlap between the door and the main body of the igloo, and will need to estimate the effect of this on their figures.

If you feel that this is too challenging, the igloo could be redrawn so that the cylindrical doorway overlaps solely with the cylindrical bottom section of the igloo and not with the hemisphere.

The task can be made more directed by providing dimensions, or less directed by suggesting that the picture is of a two-person igloo and allowing students to decide on the likely dimensions.

The drawing gives some indication of the thickness of the ice so the volume of ice, the interior volume and the surface area can all be estimated.

Students may wish to consider how adults could occupy the inside shape – sitting, lying down, and so on.

Group A

Volumes of spheres and hemispheres

Task Find the volume of a tennis ball if you know the radius

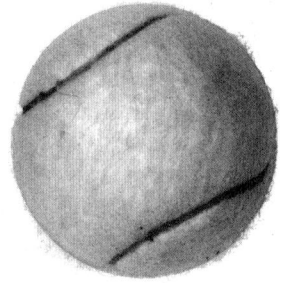

Group B

Surface areas of spheres and hemispheres

Task Find the curved surface area of a hemispherical dome if you know the radius

Group C

Volume of cylinders

Task Find the volume of a baked bean tin

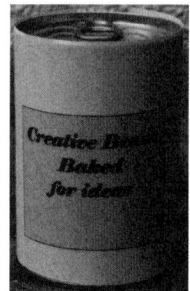

Group D

Surface area of cylinders

Task Find the total surface area of a pipe

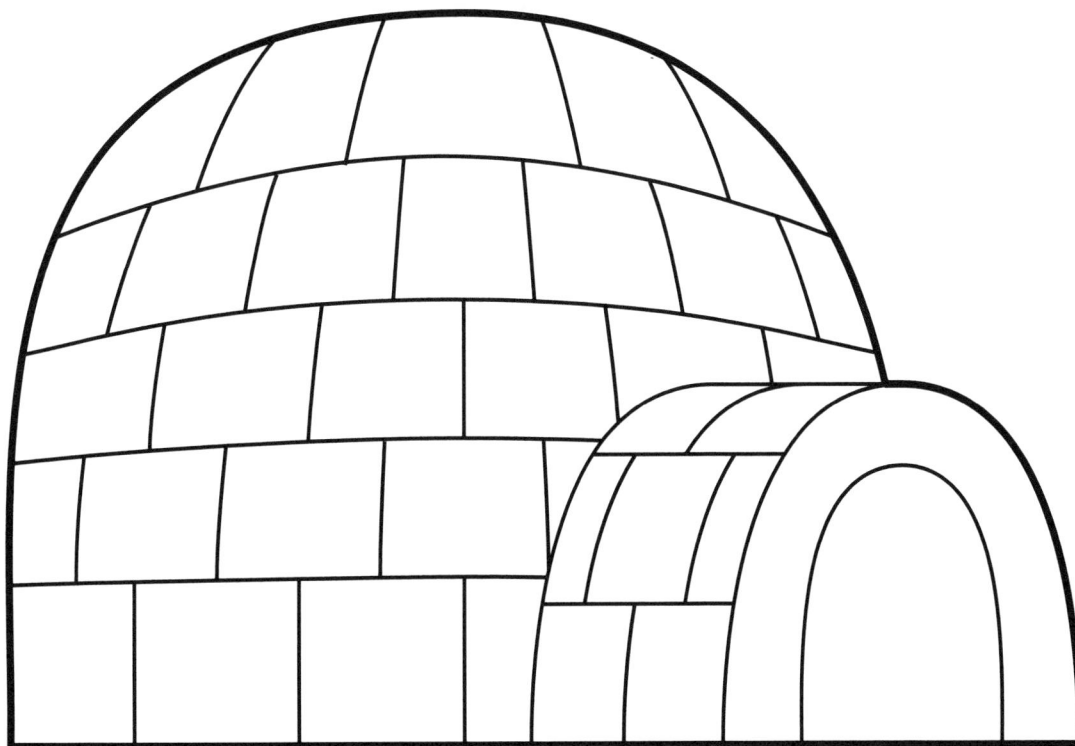

An estate agent is advertising this low-maintenance igloo as a cosy weekend retreat for two Eskimos. What details can you provide for his information sheet?

Inductive teaching with practical examples

Inductive teaching is an approach used to encourage higher-order skills in the process of sorting and categorizing data. This data may be of any form, ranging from paintings by famous artists to words with differing linguistic roots.

In most classroom situations, when students are required to sort information the task set is deductive as the categories to be used for sorting the data have already been provided by the teacher. The impression is given that there is only one correct way of approaching the task, but there may in fact be several equally valid groupings. The inductive approach opens up discussion on the common characteristics of members of a data set.

A data set is provided by the teacher or constructed by the students themselves. On the first occasion that inductive teaching is used it is likely to be more effective if data is provided; the session can then focus on the remaining stages rather than on the collection itself.

No instructions are given to the students regarding the way in which the data should be sorted – they adopt their own approach working in groups. Time needs to be given over to this initial stage of getting to know the data and preliminary sorting. Students then explain their approach to others describing the characteristics which placed certain data in the same group. If the data is numbered, feedback is more efficiently collated, a whiteboard can aid this process.

New pieces of data are then offered by the teacher, and the students have to decide if they fit in one or more of the original categories or if the data needs resorting. Data which does not fit any of the existing categories may also be offered.

The stages followed are

Presentation/collection of data

Study the data

Classify and share

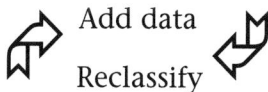

Add data

Reclassify

It is clear from this diagram that the feedback is not only a way of demonstrating present ideas but also a process in which part of the output of a system is returned to its input in order to rethink further output. The cyclical nature of the process will be easily observed in the class situation

Gradually by sorting and re-sorting, the students, encouraged through your guided questioning, will reach an understanding which allows for a hypothesis to be generated and tested and for new learning to be firmly connected to existing knowledge.

LETTERS

T | Teaching Notes

Prior knowledge

- None required although those familiar with the mathematical words may find it easier

Lesson objective

- To give pupils an understanding of the inductive approach

Suggested outcome

- Students could create their own mathematical word data set

Groupings   Pairs

Timing   Minimum 30 minutes

Equipment   None

This is a very straightforward confidence-building task to introduce the idea of inductive teaching.

Give the cards to students and ask them to sort the data in any way they choose. Give no instruction regarding sorting, no hints but explain that any arrangement is acceptable providing they can justify their choice.

They may spot the mathematical words immediately and sort accordingly, but they may not. One or two words, for example, radian, may require them to use a textbook.

The words listed are

triangle	square	angle	line	radian
vertices	degrees	ellipse	rhombus	curves
circle	centre	square units	pyramid	metres

You can add others of your own

If they choose to sort according to word length, number of vowels, or repeated consonants and so on, then provided that they are able to justify their categories all answers are acceptable. By sharing ideas they will eventually begin to talk of names and descriptors of 2D and 3D shapes, and so on.

1. raintgel	**2.** equsar	**3.** legan
4. primyad	**5.** neli	**6.** iandar
7. ceritesv	**8.** reseged	**9.** piselle
10. estrem	**11.** mosurhb	**12.** secvur
13. leccir	**14.** rentec	**15.** tinus aquers

GRAPHWORK

T Teaching Notes

Prior knowledge
Pupils need to be able to

- represent mappings expressed algebraically (Level 6)

Lesson objective
To encourage pupils to

- sketch and interpret graphs of linear, quadratic, cubic functions (Level 8)

Suggested outcome

- A poster explaining how from the equation the curve may be sketched

Groupings
Groups of 3 or 4 with similar strengths

Timing
Minimum 1 hour

Equipment
Graph paper and/or graphic calculators

These cards encourage pupils to look closely at the equations, and to decide on similarities and differences.

Students work in small groups, first looking at the cards and then deciding how the data set should be grouped. The conversations which occur during this process are crucial to the learning process and it is interesting to listen in to the decisions as they are reached.

Suggestions for questions which may be needed by some pupils at the initial stage:

- What do we mean by term?
- What do we mean by power?
- What do we mean by coefficient?

Groups then explain to others their approach and respond to questions from yourself and from other students. The students may decide to rearrange their data at this stage.

You will be making a decision at this stage as to the new data to be supplied and further discussion then takes place as to whether the existing categories can cater for this new data. Data which does not fit may also be offered.

The students will inevitably categorize finally according to the power of x, its coefficient and the constant term, but at this point no graphs have been drawn.

The connection between these findings and the students' existing knowledge of graphs should be encouraged and it may then be possible for the students to sketch graphs having plotted

$$y = x^2 \text{ and } y = x^3$$

Graphic calculators or graph plotting software provide an alternative to pencil and paper plotting which some pupils find inhibiting.

1. $y = x^2$	2. $y = x^2 + 1$	3. $y = x^2 + 2$
4. $y = x^2 - 3$	5. $y = x^3 + 4$	6. $y = -x^2 - 1$
7. $y = -x^2$	8. $y = -x^3 - 2$	9. $y = -x^3 - 4$
10. $y = 2x^2$	11. $y = 3x^3$	12. $y = 4x^2 - 1$
13. $y = 3x^2$	14. $y = 2x^2 + 1$	15. $y = -2x^2$

DATT with practical examples

De Bono (1992) has developed ten direct attention thinking tools which, as the wording suggests, direct one's thinking in a particular way. These are:

Tool 1   Consequences and Sequels
Tool 2   Plus, Minus, Interesting
Tool 3   Recognize, Analyse, Divide
Tool 4   Consider All Factors
Tool 5   Aims, Goals, Objectives
Tool 6   Alternatives, Possibilities, Choices
Tool 7   Other People's Views
Tool 8   Key Values Involved
Tool 9   First Important Priorities
Tool 10   Design/Decision, Outcome, Channels, Action

Adolescents may be particularly fixed in their way of thinking and may stubbornly refuse to change position. These tools provide you, as a teacher, with a way of encouraging your students to look at a problem from a different point of view without any loss of face!

One scenario can give rise to different areas for discussion. If, for example, students were asked to undertake a PMI (Plus, Minus, Interesting) on changing the school day to continental timing, the list would have few similarities to that using Tool 6 – Alternatives, Possibilities, Choices. A plus might be a long afternoon, a minus the early start and interesting that bus timetables would have to change. With Tool 6, however, the students might come up with various adaptations to their schooling: perhaps no attendance at school, perhaps three long school days and a four-day weekend!

The tool which I find is the most useful when thinking mathematically is CAF – Consider All Factors. It allows for a mix of numerical and qualitative data to be evaluated with equal emphasis on the soft information.

This is after all what we do on a regular basis. We make a judgement as to whether any extra cost in buying superior ingredients will result in improved taste in the prepared dish. We look at the bar charts providing weather information as part of our decision as to whether or not to book the holiday.

Examples follow of how both CAF (A Porsche or not) and PMI (The taxman sees all) may be used.

A PORSCHE OR NOT

T | Teaching Notes

Prior knowledge

Pupils need to be able to

- know the rough metric equivalents of imperial units still in daily use and convert one metric unit to another (Level 5)

- understand and use the equivalences between fractions, decimals and percentages and calculate using ratios in appropriate situations (Level 6)

Lesson objective

To encourage pupils to

- use mathematical language and symbols effectively in presenting a convincing reasoned argument including mathematical justifications (exceptional performance)

Suggested outcome

- The argument can be presented orally in the style of television motoring programmes such as *Top Gear*.

Groupings — Groups of 3 or 4

Timing — Minimum 2 hours

Equipment — Most students will wish to include easily obtainable information which is not made available in the question. Much can be obtained by asking questions of an adult. Access to the Internet is useful but not essential.

The example requires students to make a decision regarding which car to buy. This will involve judgements to be made on the picture created by numerical calculation and softer data. There is no one correct decision but the argument to support any conclusions reached must be well founded. It is important to tell the students exactly how much time they have to reach a decision: the more time they have the greater complexity they will introduce. You will need to explain the use of the CAF thinking tool.

The factors are presented to the students as separate strips. It is preferable to have the students working in pairs or small groups as there is a considerable amount of information to absorb.

The first stage is time spent in absorbing the data and then thinking about what additional information might be required to reach any decision. The running costs may be estimated if the cost of petrol is obtained, and there are factors such as road tax which have been deliberately omitted. Some students will decide to research these aspects, others will not; their reasons are important.

Suggestions for questions which may be needed by some pupils at the initial stage:

- Think about the last time your family booked a holiday.
- What were the things that mattered to your parents/guardians?
- What sort of things mattered to you?

Not all the factors are money based, such as room for sporting equipment and the 'need for speed', but the script for the television show must provide evidence of having considered all factors.

If an extension task is required then you can encourage the students to include an aspect not previously brought into the argument. For example:

Find out about Carbon Credits. Would thinking on a global scale make a difference?

A Porsche or not (Consider all Factors)

Joshua has decided to spend a significant proportion of his £135,000 lottery win on a car, the remainder he intends to invest in a building society account. Should he buy a Porsche? Why or why not?

Joshua is 45 years old

Joshua has a wife Sophie and a son Alex aged 16 who live with him

Joshua has two grown up daughters and a son-in-law who visit frequently

Joshua likes fast cars

Joshua likes to play golf

Joshua and his son both snowboard

Joshua and his son both enjoy watersports

The family already owns a small hatchback which does 10,000 miles a year

The family drives about 40,000 miles a year, some of it abroad

A Porsche costs £90,000

The Porsche depreciates at the rate of 8 per cent per year

The Porsche does 20 miles per gallon

Boot space for the Porsche is 50 litres

Porsche back seats are tiny

The top speed of the Porsche is 190 mph

Insurance for the Porsche would be £1,500 a year

An Audi costs £45,000

The Audi depreciates at the rate of 17 per cent per year

The Audi does 18 miles per gallon

Boot space for the Audi is 450 litres

The Audi seats 5

The top speed of the Audi is 150 mph

Insurance for the Audi would be £1,050 a year

THE TAXMAN SEES ALL

T   Teaching Notes

Prior knowledge

Pupils need to be able to

- interpret graphs and diagrams, including pie charts, and draw conclusions (Level 5)

- understand and use the equivalences between fractions, decimals and percentages and calculate using ratios in appropriate situations (Level 6)

Lesson objective

To encourage pupils to

- give reasons for the choices they make when investigating within mathematics and explain why particular lines of enquiry or procedures are followed and others rejected (exceptional performance)

Suggested outcome

- A debate between those who feel that the pluses outweigh the minuses and those who disagree.

Groupings

Groups of 3 or 4

Timing

Minimum 2 hours

Equipment

Most students will wish to include easily obtainable information which is not made available in the question. Much can be obtained by asking questions of an adult but access to the Internet is useful for the detail of the proposals.

It may be necessary for you to facilitate an initial discussion regarding the costs involved in running a car, as not all students will be knowledgeable about the details.

Suggestions for questions which may be needed by some pupils at the initial stage:

- When you buy a car what sort of things do you have to budget for as well as the purchase price?

The proposed charges are available on the Internet and the pie chart gives some help as to how to approach the problem. The total number of miles represented has been deliberately omitted as this forces consideration of magnitude as well as distribution.

Petrol costs need to be taken into account and other factors such as environmental issues may be included before decisions can me made about pluses and minuses.

The taxman sees all (Plus, Minus, Interesting)

The British government is considering plans to charge drivers by the mile replacing the current system of road tax and petrol duty.

George who is a salesman driving 30,000 miles a year thinks it's a bad idea, Jill who works from home but likes to travel at weekends thinks it's a really good idea.

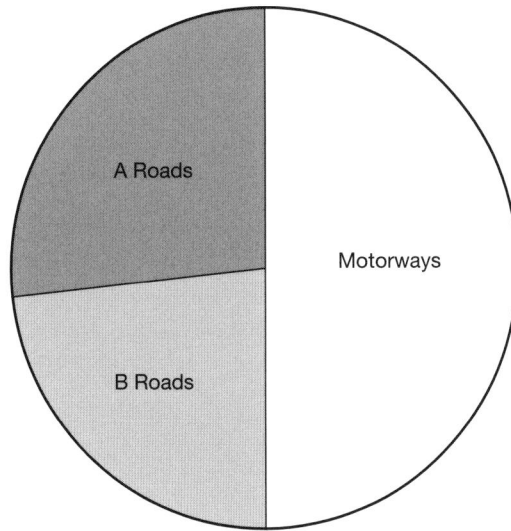

This diagram represents Jack's driving in one year.

What are the arguments for and against the pay-as-you-go approach?

What is interesting about the pay-as-you-go system?

You do not need to include George, Jack or Jill in your argument if you prefer not to, but your group must give clear mathematical evidence of both the positive and the negative.

The use of ICT

This section is looking at the use of ICT from a student perspective. I am aware of the exciting opportunities offered by whiteboards and associated software but I am also aware that, as yet, relatively few classrooms would have regular access to such facilities.

There is no doubt that the majority of students use ICT extensively both for leisure and as a tool to assist with their education. By the age of 11 most able pupils are comfortable with the use of software such as Word, Excel and PowerPoint. The Internet is a valuable source of information but students still need direction if they are not to waste hours in fruitless searches.

Many of the activities throughout this book require students to use research skills to discover facts. Many tasks require a small amount of additional information but only to a limited extent, so searches need to be time restricted with the emphasis on the use of the information obtained rather than its acquisition. Books would always be sufficient if the Internet was not available.

However, ICT has significant value as a tool to reduce the tedium of repetitive tasks which may be so demotivating to able pupils. Spreadsheets can be used to undertake various mathematical tasks and in two of the examples provided are designed to speed up repetitive tasks and the efficient use of algebraic formulae.

The third example, Fitting polygons, uses Word to replace drawing with copy and paste to tessellate polygons, allowing students to focus on the fitting together rather than the drawing of shapes.

TUG OF WAR

T | Teaching Notes

Prior knowledge
Pupils need to be able to
■ use the summation function on Excel

Lesson objective
To encourage pupils
■ to develop and follow alternative approaches, reflecting on their own lines of enquiry (Level 8)

Suggested outcome
■ Having found the solution for these particular teams, students could then work with real data, school based or otherwise, and determine whether there is a strategy to employ to minimize the number of additions required.

⚥ Groupings
Pairs or individual pupils

⧗ Timing
Minimum 1 hour

✎ Equipment
Access to computers

This task requires students to realize that they cannot use all possible members, to decide who should be left out and then allocate individuals to teams. It is probably most suitable for the lower end of the age range.

This solution is rapidly obtained using summation.

Team 1	Team 2
85	79
76	109
112	87
79	85
106	98
458	458

If possible, the initial task could be followed up by students using their own data obtained directly or from databases. It might, for instance, be interesting to introduce an acceptable percentage difference in team weights or to consider rope length.

Tug of war

Sailing ships need teams of people with strong arms and upper body strength to pull on the ropes to adjust the sails. It is thought that seamen used to compete at tug of war in their free time while on board.

Obviously it is not just about weight but we are going to assume that to make for an exciting competition, the teams must be as near to equal weight as possible.

Each team must have a combined weight of not more than 500 kilogrammes and equal numbers of competitors on each side. You have only persons A–L to choose from.

Use Excel to make it easier to do the calculations.

Choose your teams!

Person	Weight in kilograms	Person	Weight in kilograms
A	85	G	87
B	76	H	112
C	72	I	109
D	68	J	85
E	79	K	98
F	106	L	79

When you have your teams, make up your own list of possible members with fictitious weights and find out if the method you used always works.

What if you have to have a fixed number of members in each team?

SPHERES

T Teaching Notes

Prior knowledge

Pupils need to be able to
- use functions on Excel
- understand and use appropriate formulae for finding circumferences and areas of circles, areas of plane rectilinear figures and volumes of cuboids when solving problems (Level 6)
- calculate lengths, areas and volumes in plane shapes and right prisms

Lesson objective

To encourage pupils to
- solve problems involving calculating with powers, roots and numbers expressed in standard form (Level 8)
- calculate volumes of spheres (exceptional performance)

Suggested outcome
- A poster which must include the spreadsheet

Groupings
Pairs or individual pupils

Timing
Minimum 2 hours

Equipment
Access to computers and the Internet

The level of difficulty offered by this task will depend not only on the student's understanding of algebraic formulae, but also on his/her ability to work with spreadsheets. This may be something you need to take into account when deciding how to group your students.

A typical approach might be the one illustrated below:

r		r^3	$4/3\Pi r^3$		
.00005	cm	1.25E–13	5.24 – 13	cm^3	pollen grain
1.9	cm	6.86	28.73	cm^3	golf
3.28	cm	35.29	147.81	cm^3	tennis
1737.4	km	5.24E+09	2.20E+10	km^3	moon
695500	km	3.36E+17	1.41E+18	km^3	sun

By selected spherical objects as large as the sun and as small as a pollen grain you can use this opportunity to consider appropriate levels of accuracy and the use of standard form notation.

Spheres

How would you describe a sphere to a person who had never seen or touched one?

Use a spreadsheet to find the volume of a marble. You can estimate the radius of a marble or find out what it might be.

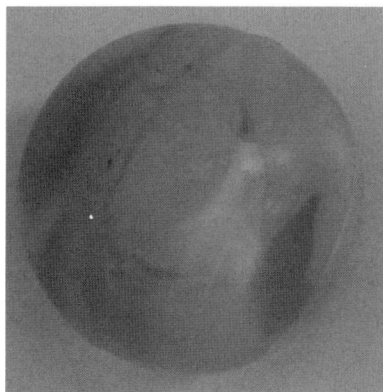

Develop your spreadsheet to find the volume of a golf ball, tennis ball, the moon, a football, the sun and a pollen grain. You will need to find out the radii.

Your spreadsheet has to provide answers which look sensible so you may have to look at different ways of writing numbers.

Extension Redo your spreadsheet so that it works in reverse and you can find the radius of any sphere if you know the volume.

POLYGONS

T | Teaching Notes

Prior knowledge
Pupils need to be able to

- use the copy and paste function on Word

- understand basic transformations

Lesson objective
To encourage pupils

- to use ICT as a tool

Suggested outcome

- A display for the classroom perhaps tiling the windows with polygons printed on OHTs

Groupings   Pairs or individual pupils

Timing   Minimum 2 hours

Equipment   Access to computers and the Internet

Using Word tessellations may be tried out quickly and accurately. Students may need additional clarification as to what is meant by sem-regular tessellations so that diagrams similar to that on the left are obtained rather than the one on the right produced by one student!

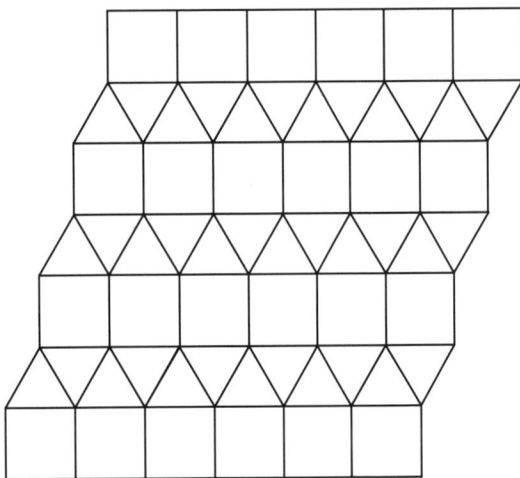

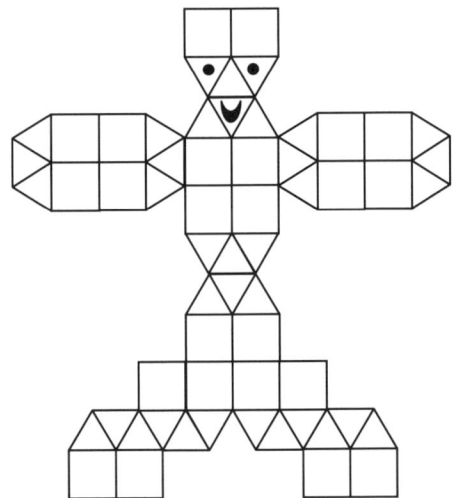

Word allows for the speed necessary to give the more able mathematician time to consider the mathematics in use rather than be hindered by the need to physically draw the shapes. I am sure that you will know of students who would find the process of drawing too inhibiting to begin.

Tessellate four small squares to make a larger square.

What you have drawn we can describe as a 4.4.4.4 tessellation: it has 4 polygons each with four sides joining at one point (a vertex).

This is a 6.6.6:

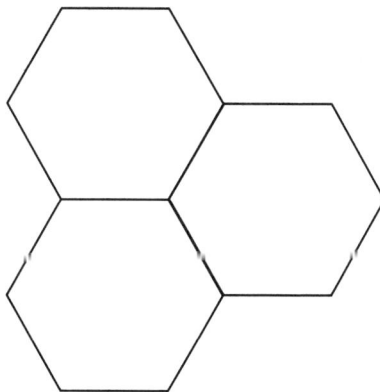

Semi-regular tessellations made up of more than one type of regular polygons with the arrangement at every vertex point being the same.

Here are some examples:

   3.3.3.4.4   3.4.6.4

Can you draw them and use Word or a paint program to show how they tessellate?

Can you find others?

Design a quilt using regular and semi-regular tesselations.

Chapter 5

Just a beginning: activities to start things off

This chapter will provide tasks which may be used as the first stage in encouraging students to take more control over their learning.

Tasks are:

- Trick or treat

- Number bases

- Probable or what?

- Hats

- Garden maths

- Crowds.

Many researchers have shown how the type and quantity of information provided, and the way in which it is presented to students, has an effect on the creative response (Runco and Sakamoto in Sternberg, 1999: 80). The worksheets are designed to capture the imagination of the students, but for this initial stage in 'releasing the brake' the problems are laid out such that attention and interest are fostered while providing the reassurance of some structure. The opportunity for students to find creative solutions always exists and should be encouraged.

Students require little new knowledge to achieve success and any information needed is readily accessible; any specific sources are listed in the teaching notes. The thought processes of your students will be clearly visible to you, as a teacher, and many tasks expect the students to feed back to others in the class.

As with tasks in Chapter 4, the way in which findings are presented will depend upon the teacher and on student preferences but suggestions are given for possible outcomes. Many pupils if given total freedom will choose to use PowerPoint to present to peers and, as always, a balance needs to be struck between time spent on the mathematics and time spent creating the presentation.

Suggestions for outcomes are provided in the teaching notes to be read alongside each task, but there are numerous other ways in which any of the findings associated with any one activity might be presented. The options offered tend to be those less frequently used within a mathematics classroom but are not uncommon in other curriculum areas.

My own research showed very clearly how highly students value being given the chance to exercise control over their work both in terms of direction and outcome. Some students prefer to work alone, others in groups; the decision as to whether freedom of choice is the best promoter of learning must rest with you as the teacher.

Suggestions are given in the teachers' notes for group sizes but these are not in any way prescriptive: the class and the individuals within will determine the group structure.

Whether you decide that the students should work as groups or not, you need to provide ample opportunity for discussion to clarify thinking, to reflect on the possibilities and to maximize the learning. Your knowledge of your students is vitally important.

Some research has suggested that highly creative teachers can actually have a negative effect on the creativity of their pupils. You need to be aware of this and not take over the task by offering ideas (however exciting!) rather than allowing the students to develop their own.

TRICK OR TREAT

T | **Teaching Notes**

Prior knowledge

Pupils need to be able to

- solve numerical problems involving multiplication and division with numbers of any size, using a calculator efficiently and appropriately (Level 7)

- appreciate the imprecision of measurement (Level 7)

- understand and use compound measure, such as speed (Level 7)

Lesson objective

To encourage pupils to

- give reasons for the choices they make when investigating within mathematics and explain why particular lines of enquiry or procedures are followed and others rejected (exceptional performance)

- apply the mathematics they know in familiar and unfamiliar contexts (exceptional performance)

Suggested outcome

- A poster illustrating possible routes

- A script for a news report of the 'expedition'

Groupings   Small groups of 3 or 4 with similar strengths

Timing   Minimum 2 hours

Equipment   Stopwatches, thread, A3 paper

The map provided is deliberately not easy to work with in that the roads are not straight. You could design a simpler version if you preferred a more structured response or the task could be made more realistic by using a map of the local area.

The task therefore involves estimating distances, walking speeds and time spent at each house and then deciding on how to optimize the gains from two hours of trick or treating. The conclusions reached by each group will not be the same. This does not matter as it is the quality of the thinking rather than the solution itself which is important. Pupils must be told at the beginning of the activity that there is no single correct answer.

Trick or treat

Joe and his friends decide to go trick or treating but only to houses where they know the occupants. These houses are marked with numbers on the map. They start from Joe's house which is at the junction of Eastern Way and Leighton Road.

The distance between houses 4 and 5 is 0.3 miles. They are likely to get similar treats from each of the houses but every visit takes time and they have to be back at Joe's in two hours for a barbeque supper. Should they visit every house? What do you think is their optimum route?

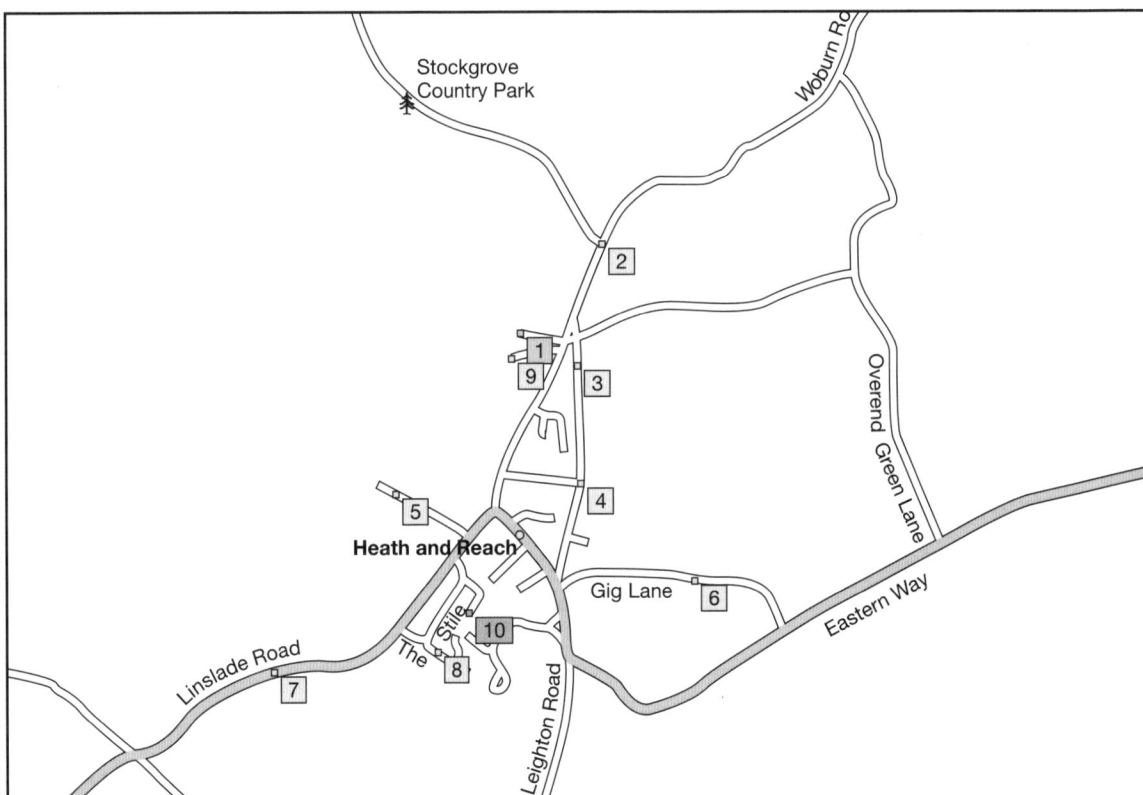

Extension Select one of the house numbers you plan to visit at random. How would you change your route if you knew that the owners of this house were out?

Pupils are likely to adopt a trial-and-error approach to route-finding. If exceptional mathematicians show a deep interest, they may wish to look at AS (advanced subsidiary) discrete mathematics textbooks looking at, for example, Travelling Salesman problems.

As explained in the introduction to this chapter, this is the first stage of 'letting go'. Although this task is well defined, students have to make decisions on numerical values which then impact on the choices made. This is in itself a big change for those pupils who previously have worked only from textbooks which provide all necessary data and precise answers.

Suggestions for questions which may be needed by some pupils at the initial stage:

- How can you find out how far it is between the houses?
- How do you know how long it will take to walk?
- How will you estimate the time spent at each house?

NUMBER BASES

T | Teaching Notes

Prior knowledge

Pupils need to be able to

- understand basic mathematical rules of addition, subtraction, multiplication and division

- solve problems involving calculating with powers (Level 8)

Lesson objective

To encourage pupils to

- apply the mathematics they know in familiar and unfamiliar contexts (exceptional performance)

Suggested outcome

- A debate to consider: 'Base 10 is best'

👤 Groupings | Small groups of 2 or 3 with similar strengths

⧗ Timing | Minimum 2 hours

✎ Equipment | Not essential but a textbook with exercises on number bases might be useful

This task will need very little teacher intervention in the early stages but you will need to be involved in discussions as the task progresses.

Able mathematicians will quickly understand the principles and will enjoy playing with the ideas. Number bases used to be included as part of the curriculum, so materials to support any follow-on activities such as multiplication or division are likely to be available.

Suggestions for questions which may be needed by some pupils at the initial stage:

- How many pennies were there in an old shilling?
- How many old shillings in a pound?
- Why do you think we count in tens, hundreds etc?

Students may decide to use other alphabets as numerical symbols or the Roman numeral system; they may use shapes with differing numbers of sides.

For example in base 15
28 in base 10 could be 1Δ if a triangle is used for 13, ☐ for 14, and so on.

They may decide to use a pattern of dots and dashes or even musical notes.

Having devised a system, students will find some more workable than others, especially if you encourage them to consider fractional values. It may then lead them into discussions regarding exact values.

Number bases

In our culture we generally use 10 as the base for our number system so that

$$293 \text{ represents } 2 \times 10^2 + 9 \times 10 + 3 \times 1$$

But if we used base 2, 293 would be written as 100100101, that is

$$\mathbf{1} \times 2^8 + \mathbf{0} \times 2^7 + \mathbf{0} \times 2^6 + \mathbf{1} \times 2^5 + \mathbf{0} \times 2^4 + \mathbf{0} \times 2^3 + \mathbf{1} \times 2^2 + \mathbf{0} \times 2^1 + \mathbf{1}$$

If we use bases less than or equal to 10 we have all the symbols we need, but for base 11 we need a symbol for 10 (T) and for base 12 a symbol for 10 and 11 (T and E).

In base 12, 3TE + T9E = 128T. Check this for yourself.

Decide on a base bigger than 12 that you would like to work in and devise your own symbols for the 'digits' you need. The bigger the base the more symbols are required – hexa-decimal (base 16) often uses a–f but you might have more exciting ideas. The symbols do not have to be from existing lists, you can create your own, but they do have to make some sort of sense by suggesting an increase of one each time.

Choose four base 10 numbers bigger than 293 and change them to your base.

Can you add your numbers together in the base you are working in? What about subtraction?

You can check your answers by changing the numbers back to base 10.

Have you used only whole numbers? What would happen if you needed to change fractional numbers to your base?

Extension Could you easily recognize a leap year working in your base?

PROBABLE OR WHAT?

T Teaching Notes

Prior knowledge

Pupils need to be able to

- use their knowledge that the total probability of all the mutually exclusive outcomes of an experiment is 1 (Level 6)

- understand the effects of multiplying and dividing by numbers between 0 and 1 (Level 7)

- understand how to calculate the probability of a compound event and use this in solving problems (Level 8)

Lesson objective

To encourage pupils to

- recognize when and how to work with probabilities associated with independent mutually exclusive events (exceptional performance)

- apply the mathematics they know in familiar and unfamiliar contexts (exceptional performance)

Suggested outcome

- The longest chain of fractions with associated events

Groupings

Small groups of 3 or 4 with similar strengths

Timing

Minimum 1 hour

Equipment

- Information on independent events

- Coins/dice/cards for initial demonstration if needed

This task brings together an understanding of probability and the effective use of fractions with the option to introduce a competitive element.

There are numerous possibilities a student might list basing probabilities on events such as an hour of the day, heads/tails, primes on a die, suits from a pack of cards, digits, and so on.

Suggestions for questions which may be needed by some pupils at the initial stage:

- Is it more likely or less likely for two events to occur at the same time?
- How do you think we might find the probability?

Probable or what?

Two independent events have a combined probability of them both happening of $\frac{1}{24}$.

What does the word 'independent' mean?

List as many pairs of probabilities as possible which multiply to give $\frac{1}{24}$, describing events which would match individual probabilities.

For example: $\frac{1}{8} \times \frac{1}{3}$

Probability of 3 boys in a family of three children × Probability of getting a multiple of 3 on a die.

Try to avoid constructing situations such as placing the appropriate number of differently coloured beads in a bag, making events easier to find. Try instead to use dice, cards, coins, calendars, and so on.

What if there were three independent results with a probability of all three happening of $\frac{1}{24}$?

How many different combinations of three fractions less than 1, can you find that multiply to give $\frac{1}{24}$?

Can you give examples of possible events for the trios as you did with the pairs?

How many independent events can you get up to still keeping the overall probability as $\frac{1}{24}$?

The process could start simply with fractions such as in the example such as $\frac{1}{8} \times \frac{1}{3}$ (Probability of 3 heads when 3 coins are tossed) × (Probability of a multiple of 3 when a die is thrown) and develop to include more events:

$$\frac{1}{2} \times \frac{3}{4} \times \frac{1}{3} \times \frac{1}{2} \times \frac{2}{3}$$

(Probability of a head when one coin is tossed) × (Probability of heart, club or diamond from a pack of cards) × (Probability of a number bigger than 4 when a die is tossed) × (Probability of getting a multiple of 2 on a die) × (Probability of the second hand on a stopwatch being between 0 and 40).

The chain of fractions gets less easy as it becomes longer, especially with the constraints imposed on the situations.

$$\frac{7}{10} \times \frac{5}{8} \times \frac{1}{2} \times \frac{2}{3} \times \frac{1}{2} \times \frac{4}{5} \times \frac{5}{7}$$

Calculators could be used to check multiplication strings.

HATS

T Teaching Notes

Prior knowledge
Pupils need to be able to

■ understand and use appropriate formulae for finding circumferences and areas of circles, areas of plane rectilinear figures and volumes of cuboids when solving problems (Level 6)

Lesson objective
To encourage pupils to

■ apply the mathematics they know in familiar and unfamiliar contexts (exceptional performance)

Suggested outcomes

■ A presentation of their findings to other groups

■ Instructions without diagrams to construct hats could be tried out by a younger age group

| | Groupings | Small groups of 2 or 3 with similar strengths |

Groupings Small groups of 2 or 3 with similar strengths

Timing Minimum 1 hour 30 minutes

Equipment

■ Access to the Internet or library

■ Paper

■ Scissors

It is worth noting that being an able mathematician does not necessarily imply dexterity and that, for some, the folding will be an issue.

As with most of the tasks, this activity may be undertaken either individually or in small groups. If you decide on groups, the ongoing discussion leads to enhanced understanding and promotes learning, but it is worth remembering that it is likely that most students will wish to make their own hat.

The decision over how easy or difficult the hat design is rests with you as the class teacher, but it is important that the paper supplied does not allow a hat of the given size to be made initially.

The assumption often made is that able students will easily be able to make a hat, but in my experience this is not necessarily the case. My advice would be to offer a simple design and allow the students themselves to add the complexity.

There will be many answers to the surface area depending on the original design and how the students interpret the folds, but the importance is on the explanation of how they have applied existing skills.

When scaling up the hat, there will be decisions to be made about where the contact is between hat and head. It is depth of consideration and the quality of the debate that are of importance.

The video sequence demonstrating paper-folding by a robot will raise awareness of the complexity. If diagrams are permitted, then producing instructions is much more straightforward. If you and your students decide that diagrams are not allowed, then there will be a need for precise mathematical language.

Students previously undertaking this task have extended work on hats to include:

- using prints of Dr Seuss's Cat-in-a-hat to work out the empty space above the cat's head
- making 'crowns' which tessellate, as given out by burger restaurants.

Your teacher may have the instructions for you or you may already know how to make a folded paper hat. Otherwise look on the Internet or in the library to find instructions. It can be any design but must use a single sheet of paper without cutting. Make a hat of any size.

Find the surface area of your finished hat. Be ready to explain to your teacher and other groups how you worked this out.

What size paper would you need to have started with to make a hat for a person with a head circumference of 53 cms? How have you reached this decision?

Don't forget your group may have to feed back to the rest of the class. Prepare what you are going to say.

Robots have been designed to produce origami hats but need clear instructions. Look at www-2.cs.cmu.edu/%7Edevin/folder1.mov

Write instructions so that a machine could make your hat.

GARDEN MATHS

T Teaching Notes

Prior knowledge

Pupils need to be able to

- understand and use appropriate formulae for finding circumferences and areas of circles, areas of plane rectilinear figures (Level 6)

- determine the locus of an object moving according to a rule (Level 7)

Lesson objective

To encourage pupils

- to calculate lengths of circular arcs and areas of sectors

Suggested outcome

- Groups could demonstrate using their plans as supporting evidence.

Groupings Small groups of 3 or 4 with similar strengths

Timing Minimum 2 hours

Equipment A3 paper, squared paper, OHTs

The first part of this task is highly structured but still gives opportunity for the students to make choices and to be creative within that structure. They may need estimated measurements for gardens around their home or school before beginning the task.

Suggestions for questions which may be needed by some pupils at the initial stage:

- Think of a garden you have walked through recently.
- What shapes were the flower beds?
- What would you estimate the diameter of a tree trunk to be?

Students will need to consider how long a piece of rope is sensible for the goat(s) and how wide a mower blade is. They will need to decide how to record and perhaps measure the overlap in each case. Photocopies of the garden design or the use of tracing paper or OHT overlays would avoid confusing diagrams and the irritation which might then follow.

The students may decide to change their designs for the garden when munching/mowing becomes difficult.

Garden maths

Read through everything on this page before starting your work. If you do not do so you may find that you have to redo some of the early stages.

Draw a plan of a rectangular garden for which the ratio of length to width is at least 2:1. Make it clear what scale you are using.

Decide where you would like to place two trees and surround the trunks with a narrow circular bed. Add two flower beds which must be different mathematical shapes (not rectangular) each of area 4 square metres, and a circular pond of radius 1 metre. Add a path of width 0.5 metres around the pond.

The remaining area is grassed. What is the area of the lawn in your garden?

The owner of this garden has decided that she will keep a goat to eat the grass. Where would you tether the goat? The rope must not be so short that the goat feels unable to move or too long so that it gets tangled up. Depending on the size of your garden, there may be more than one position to reach all the grass, especially as the goat is best kept away from the flower beds.

Show the position(s) clearly on your plan. Are some areas of grass going to be shorter than others? Will some be left uncut? Neither is desirable.

Extension Would a mower be able to go over each area just once or would some parts of the lawn have more than one cut?

CROWDS

T   Teaching Notes

Prior knowledge
Pupils need to be able to

- specify and test hypotheses (Level 7)

- determine the modal class and estimate the mean, median and range of sets of grouped data (Level 7)

- use measures of average and range (Level 7)

Lesson objective
To encourage pupils to

- understand how different methods of sampling and different sample sizes may affect the reliability of conclusions drawn (exceptional performance)

Suggested outcome

- Take one example of a reported demonstration and produce an argument agreeing or disagreeing with the estimate

Groupings
Small groups of 3 or 4 with similar strengths

Timing
Minimum 1 hour

Equipment

- Some photographs of crowds

- Street maps

- Access to the Internet

- Measuring equipment

I would expect students to suggest that they would need a map of the area and that they would indicate on the map the area covered by the demonstration at any instant in time. The task of finding the area of ground covered by the crowd will vary in difficulty. You will need to choose your crowd to match the ability of your students and the maps that are accessible online.

The students will need to estimate the space required per individual, and this is likely to develop into a practical activity involving some sampling and some consideration of mean and range.

Suggestions for questions which may be needed by some pupils at the initial stage:

- How could the headteacher estimate school attendance from a school assembly?
- How do promoters know how many tickets to sell for gigs in places like Hyde Park?

Discussion could consider how accurate an estimate needs to be to be valid in relation to the size of the crowd.

Crowds

Because tickets are sold for gigs, football matches, and so on, it is easy to estimate the size of a crowd.

When people turn up to welcome a successful rugby team home or to demonstrate against something they feel strongly about it is less easy to estimate numbers.

In Rome, two and a half million people were said to be demonstrating against war, in London one and three-quarter million and eighty thousand in Amsterdam. You can find photographs of crowds on the Internet.

Imagine that you were a reporter watching an event as it happened. How would you obtain an estimate of the number attending? Television reporters seem to be able to decide on a number quite quickly. How accurate do you think they are?

Chapter 6

Getting more confident

This chapter will provide tasks which may be used as the second stage in encouraging students to take more control over their learning and therefore the tasks are less structured than those in Chapter 5.

Tasks are:

■ One poem

■ Oh Grandma!

■ Shall we dance?

■ Beauty.

The activities in the previous chapter provide a degree of choice but the tasks are structured in a way which determines directions most likely to be followed by the students. In this chapter the level of control is lessened but questions are still put to the student so as to direct the learning: in Chapter 7 the choice of direction rests solely with the learner.

By undertaking tasks from both chapters, students, as recommended by Eyre, will be encouraged to develop their skills as independent learners by:

– *organising their own work*
– *carrying out unaided tasks which stretch their capabilities*
– *making choices about their work*
– *developing the ability to evaluate their work and so become self critical.*
  (Eyre, 1997: 147)

These tasks move further away from those routinely set in mathematics and there may be some bewilderment initially. These feelings may be shared by teacher and student but will be only briefly experienced especially if examples from earlier chapters have been used to support the change.

The graphs below reflect the opinions expressed by two groups of around 30 Year 9 students on being given the freedom to select an approach, in terms of the mathematics, and to decide whether they worked as groups or individuals. The questionnaire was used on three separate occasions and the results collated.

The first graph shows how the two groups (A and B) felt about being able to select their own way of working on three different creative tasks.

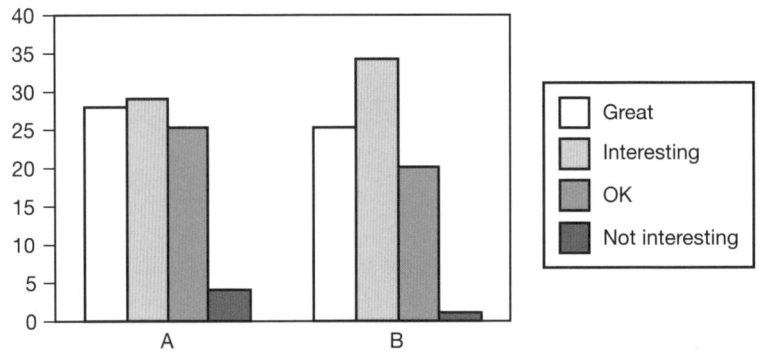

The second graph clearly indicates their strong preference for being able to choose to work in groups.

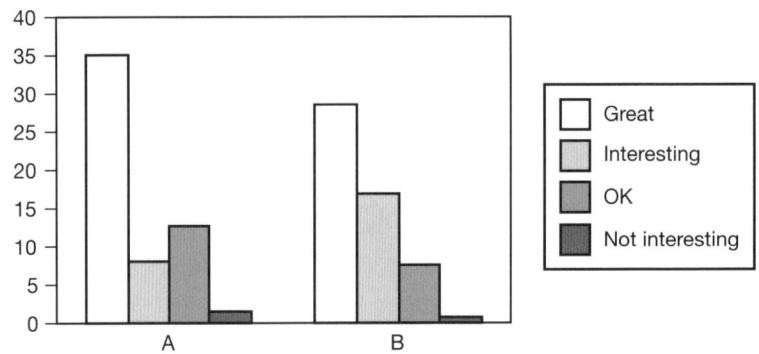

While all activities in this section are possible for individual students to undertake, you will find that group work allows for a more supportive learning environment for those who are less certain and encourages greater risk-taking.

You will notice that the worksheets have less detail. This is to encourage ideas beyond that normally expected within mathematics classrooms. While there is no need for you as a teacher to work through any particular task it is beneficial to allow yourself a little time to think about the possibilities.

ONE POEM

T | Teaching Notes

Prior knowledge

Pupils (may) need to be able to

- solve numerical problems involving multiplication and division with numbers of any size, using a calculator efficiently and appropriately (Level 7)

- appreciate the imprecision of measurement (Level 7)

- understand and use proportional changes (Level 7)

- understand and use compound measure, such as speed (Level 7)

- calculate lengths, areas and volumes in plane shapes and right prisms (Level 7)

- understand and apply Pythagoras' theorem when solving problems in two dimensions (Level 7)

Because of the nature of this task it is not possible to predict exactly what prior knowledge the students needs.

Lesson objective

To encourage pupils to

- give reasons for the choices they make when investigating within mathematics and explain why particular lines of enquiry or procedures are followed and others rejected (exceptional performance)

- apply the mathematics they know in familiar and unfamiliar contexts (exceptional performance)

Suggested outcome

- A presentation to another group of students to clarify some of the complex decision-making involved

Groupings

Small groups of 3 or 4 with similar strengths

Timing

Minimum 3 hours

Equipment

Internet access or alternative research facilities

The questions set give this task some limited structure but the stimulus is unusual and the questions deliberately ambiguous so that genuine creative responses may be encouraged.

There are no numbers in the poem so there is a need for initial research to obtain data regarding the world population and the 'size' of the seas. There could be a debate regarding male/female/child populations.

Suggestions for questions which may be needed by some pupils at the initial stage:

■ How can we find out about the seas?
■ What do we know about the surface area?
■ What do we know about the depth?

Gifted students are able to deal with ambiguity and should therefore be able to make a decision as to whether they think that for seas the appropriate measure is volume or surface area, especially when these are to be added. Your role is to encourage the discussion, not to offer an opinion.

To come to any conclusions the decision then has to be made about the shape, including depth if volume has been used, of the 'one' sea, what is meant by 'across' and how/if swimming speed relates to height. Students will need your encouragement to consider these complex issues.

As regards adding people together, the likelihood is that they would use scale factors, for example eight people would create an enlargement scale factor 2 of the original, and so on. By cube rooting the world's population some estimate of the size of the man could be obtained but they may decide on other approaches.

The process involves many decisions for which there is obviously no definitive answer but which raise issues that are relevant when working with algebraic functions.

One poem

Do you know this nursery rhyme? Have you ever thought about it?

If all the seas were one sea,

What a great sea that would be!

If all the trees were one tree,

What a great tree that would be!

If all the axes were one axe,

What a great axe that would be!

If all the men were one man,

What a great man he would be!

And if the great man took the great axe,

And cut down the great tree,

And let it fall into the great sea,

What a great splash-splash that would be!

How big do you think the sea would be?

What about the man, the tree or the axe?

Extension How long would it take the one man to swim across the one sea?

OH GRANDMA!

T | Teaching Notes

Prior knowledge
- None required for the initial task
- Students need to be able to calculate with powers, roots and numbers expressed in standard form (Level 8) for the extension

Lesson objective
To encourage pupils to
- apply the mathematics they know in familiar and unfamiliar contexts (exceptional performance)

Suggested outcome
- Brief verbal feedback and discussion of ideas

Groupings   This could be a whole-class activity for the initial task

Timing   Minimum 30 minutes

Equipment
- Copies of *Little Red Riding Hood* and *Three Little Pigs*
- Access to the Internet or A level textbooks required to explore the topic in more depth

This is a short activity which could be included in a lesson on, for example, proportion.

The story gives many contradictory statements regarding the size of the wolf, and it is simply a matter of selecting the appropriate evidence; similarly for the *Three Little Pigs*.

If available to you, a whiteboard could be used to move the story into 'for' and 'against' categories.

If the students wish to explore further, then the logarithmic relationship gives them a starting point. It would be possible to generate graphs either by hand, or using technology, which could be used to predict the eye size of various pets.

Oh Grandma, what big eyes you have!

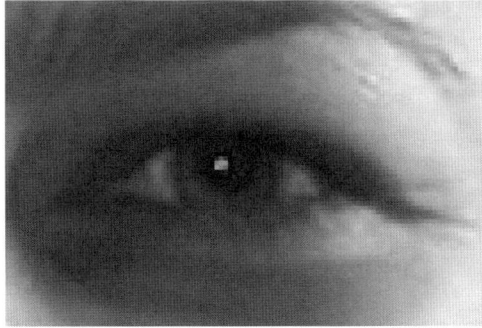

Bigger animals tend on average to have bigger eyes.

Do you think that the wolf in the story would have had bigger eyes than grandma based on other comments made by Red Riding Hood?

What other evidence can you find in the story to support your point of view?

In the *Three Little Pigs* the wolf falls down the chimney. Is he bigger or smaller than the wolf in *Little Red Riding Hood*? Why do you think that?

Extension Howard C. Howland, Stacey Merola and Jennifer R. Basarab found that there was a logarithmic relationship between animals' body weight and eye size for all vertebrates. www.innovations-report.com/html/reports/life_sciences/report-32176.html

What does logarithmic mean?

SHALL WE DANCE?

T | Teaching Notes

Prior knowledge
Pupils will need to

- be able to transform shapes

- (and possibly) devise instructions for a computer to generate and transform shapes and paths (Level 6)

Lesson objective
To encourage pupils to

- apply the mathematics they know in familiar and unfamiliar contexts (exceptional performance)

Suggested outcome

- Students could perform their dance/use a robot to create the patterns on the floor

Groupings

Pairs with similar strengths

Timing

Minimum 1 hour

Equipment

- Access to the Internet or books required to explore the topic in more depth

- Space to dance!

- Not essential but for non-dancers a computer controlled robot is useful

This activity is designed for kinaesthetic learners who will enjoy moving around as they create the mathematical instructions for their dance.

It is unlikely that there will be a need for you to prompt the students as the cues are built into the worksheet.

It is possible to extend the work to cover Euler's laws for networks if the students find the extension question absorbing.

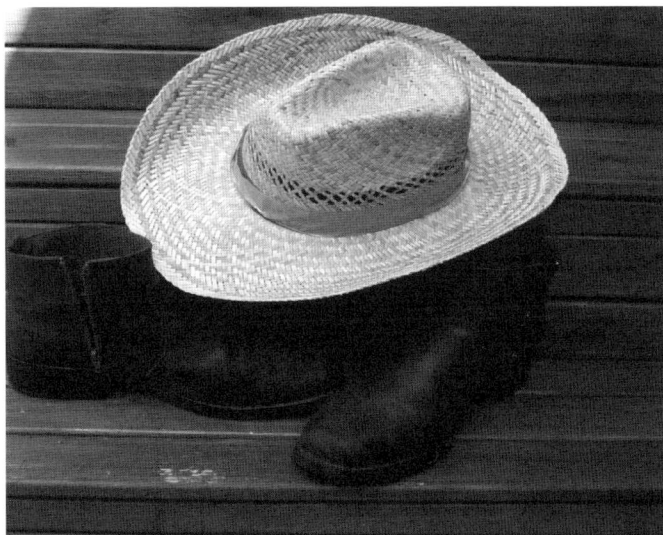

Find out about line dancing.

What mathematical transformations do the dancers demonstrate

■ on their own?

■ as pairs?

■ as a group?

Design and describe in detail your own one-minute dance routine by using co-ordinate axes and mathematical language of transformations to choreograph it. It can be any style of dance you like but you must be able to choreograph it mathematically.

Hand your plans to another group. Are they able to perform it?

Extension Think about the floor lines which are covered. Can you redraw your floor pattern without taking your pencil off the paper or without going over any line twice?

BEAUTY

T | Teaching Notes

Prior knowledge
Pupils (may) need to be able to

- refine or extend the mathematics used to generate fuller solutions (Level 7)

- justify their generalizations, arguments or solutions, showing some insight into the mathematical structure of the problem (Level 7)

Lesson objective
To encourage pupils to

- apply the mathematics they know in familiar and unfamiliar contexts (exceptional performance)

- use mathematical language and symbols effectively in presenting a convincing reasoned argument including mathematical justifications (exceptional performance)

Suggested outcome

The way in which the presentation is *constructed* has to in some way convey the findings, for example

- a PowerPoint presentation might have the number of points per screen equal to that Fibonacci number so that the third slide had 2 points, the fifth 5 points, and so on

- a poster would have rectangles of the appropriate proportions

- a write-up would have paragraphs with lengths in proportion to the golden ratio

Groupings

Small groups of 3 or 4 with similar strengths

Timing

Minimum 3 hours

Equipment

Internet access or alternative research facilities. Students will be employing research skills from the outset as they may not be aware of the links with Fibonacci and the golden ratio

You may wish to share some information with the students at the outset or you may prefer to encourage them to research the ideas themselves. They will almost certainly need a prompt to make the link.

Find out about paintings by Mondrian. They are very mathematical. Why?

The picture is of a pine cone. What is the link between Mondrian's paintings and the cone?

What has it got to do with credit cards and Virgil's Aeneid?

You are a mathematical artist. Produce a work of art which *explains* the mathematics that you have found out. Use the principles you now know to present your ideas 'beautifully'.

Suggestions for questions which may be needed by some pupils at the initial stage:

■ Do you know about the Fibonacci sequence?

In 1202 Fibonacci investigated how fast rabbits could breed. The sequence begins with a newly born pair of rabbits, one male and one female. Rabbits are capable of mating at the age of one month so that at the end of the second month the female can produce another pair of rabbits. Fibonacci assumed that the female always produced a pair of newborns and that none died.

The number of pairs at the start of each month is therefore

1   1   2   3   5   8   13   and so on

If we divide each number by the previous term then the ratio reaches a limit of 1.618 which is the golden ratio used by artists across the centuries.

The education centre at Eden in St Austell, Cornwall (see photograph) is based on the Fibonacci series and plant spirals, as in the pine cone.

Credit cards are a perfect golden rectangle.

The mathematics involved is straightforward and I would recommend this task for use with those with obvious ability but whose interest lie in the more 'creative' curriculum areas.

Chapter 7

Ideas just flow

> This chapter will provide tasks which may be used as the third stage in encouraging students to take more control over their learning and are therefore even less structured than those in earlier chapters.
>
> Tasks are:
>
> ■ Murphy's Law
>
> ■ Pyramids
>
> ■ Charlie's garden.

In this chapter the student material is designed to stimulate mathematical activity but no precise instructions are given. Students may choose to explore any avenues which are of interest, may work individually or in groups and may present their findings in any way they so choose.

These tasks were used with Year 9 students, aged 13 to 14, and the work produced was completed by the students as homework. This was not ideal but was the only way in which the school felt it could include such activities in a packed curriculum. Sixty top set students were involved, so there was a wide range of ability.

It is impossible to convey the excitement generated by the freedom to explore and express but perhaps one indicator is that I had to collect a storage crate to contain all the work of one class after just one 90-minute homework session! Neither class was taught by me, although I did introduce each task.

The first task is Murphy's Law, the second Pyramids and the third Charlie's garden. The activities were designed as part of a research project so there was no lead in through other less open tasks, but student response was excellent. They were made aware that there was no right or wrong answer and, consequently, experienced no fear of failure.

Interestingly, parents who had been informed of the activities were also very positive and commented on how much their children – all able mathematicians – had enjoyed and benefited from undertaking the tasks. Parents of these students who are now Year 13 (aged 17 to 18) still comment positively on the experience, indicating perhaps how they too value the opportunity for their sons and daughters to display the depth and breadth of their mathematical understanding.

There was a link between the activity and recently studied curriculum content which helped ideas to flow. This would not be an essential factor but in the particular circumstances was beneficial.

MURPHY'S LAW

T | Teaching Notes

Prior knowledge
Pupils need to be able to
- understand relative frequency as an estimate of probability and use this to compare outcomes of experiments (Level 7)

Lesson objective
To encourage pupils to
- give reasons for the choices they make when investigating within mathematics and explain why particular lines of enquiry or procedures are followed and others rejected (exceptional performance)

- apply the mathematics they know in familiar and unfamiliar contexts (exceptional performance)

Suggested outcomes
No suggestions are made to the students, but these are a selection of those previously created:

- Experimental work: handwritten, word processed then printed/on disk/CD-ROMs

- PowerPoint incorporating photographs and video clips, posters

- Rap

- Secret diary: each day devoted to a Murphy's Law event

- A collection of photographs: interesting photos that failed in the production and vice versa

- A website on Murphy Law that did not quite work!

Groupings | Student free choice

Timing | Minimum 1 hour 30 minutes

Equipment
- Internet access or alternative research facilities

- Possibly toast and butter (Blu-Tack preferably to protect carpets!)

Able students have little difficulty in setting themselves a task based on the information given, as indicated by their ability to work on the task with no teacher intervention.

Murphy's Law

Edward A. Murphy Jr was an engineer working on US Air Force experiments in 1949 to test human tolerance of acceleration. One experiment involved a set of 16 instruments being attached to different parts of the test subject's body. As with most things, there was a right way and a wrong way round for the sensors and guess what happened? All 16 were attached incorrectly!

Murphy's original statement was: 'If there are two or more ways to do something, and one of those ways can result in a catastrophe, then someone will do it.'

The version you hear most often is: 'If anything can go wrong it will go wrong.'

One of the statements closely linked with Murphy's law is: 'Toast always lands butter-side down' but there are lots of similar comments.

Choose an aspect of Murphy's Law to investigate. You may choose to carry out experiments or surveys. You may decide on a theoretical approach but whatever you decide to do it must include mathematical calculations.

You may work individually or in groups and present your work in whatever way you wish.

Photocopiable: Creative Maths Activities for Able Students – Ideas for Working with Children Aged 11 to 14
Paul Chapman Publishing 2006 © Anne Price

Many will be interested in the butter-side-down idea and will carry out experiments devising rules for how the toast should be dropped, from what height, and so on. The students themselves had the idea of replacing the butter with Blu-Tack to protect carpets.

Other students may, like those I worked with, choose to look at queues at checkouts or the link between waxing one's hair and the strength of the prevailing wind experienced that day!

This is the approach used by one boy:

Chosen problem: Picking somewhere on a map near a crease

The way in which he undertook the task was to:

- select a 1:50000 ordinance survey map and find the position of all the creases
- find the area of the map
- estimate the width of each crease and find the total creased area allowing for overlap
- estimate the probability
- discuss the issues arising from the decisions he had made
- convert the areas to 'real land' and consider what this probability might mean in practice.

PYRAMIDS

T Teaching Notes

Prior knowledge

Pupils need to be able to

- understand and apply Pythagoras' theorem when solving problems in two dimensions (Level 7)

- calculate lengths, areas and volumes in plane shapes and right prisms (Level 7)

- use sine, cosine and tangent in right-angled triangles when solving problems in two dimensions (Level 8)

Lesson objective

To encourage pupils to

- apply the mathematics they know in familiar and unfamiliar contexts (exceptional performance)

- (possibly) use sine, cosine and tangent of angles of any size, and Pythagoras' theorem when solving problems in two and three dimensions (exceptional performance)

Suggested outcomes

- No suggestions are made to the students but these are a selection of previous responses

- Exact scale models

- Board game using Egyptian number system

- Changing Tombs (*Changing Rooms* with a difference!)

- Website game using the pyramid as a target for a projectile

Groupings

Student free choice

Timing

Minimum 1 hour 30 minutes

Equipment

- Internet access or alternative research facilities

- Paper, glue and so on for model-making

For many gifted mathematicians the opportunity to 'cut and stick' is phased out at a very early stage of their learning. The choice offered allows students to re-establish a more kinaesthetic approach to their mathematics by producing scale models, and you are likely to find that there are several who wish to take the opportunity. The models will be made with precision and with all the necessary calculations to allow for exact measurements. There is a significant opportunity to use trigonometry to a more advanced level than the classroom permits, with students developing the use of the method in three dimensions. While most students will produce work closely related to the stimulus, some will use it as an opportunity to pursue an interest of their own, perhaps with surprising results.

The following section describes the work produced by a Year 9, 13-year-old, and demonstrates clearly how the student made the task his own in that the pyramid could in fact have been any object, although he did take into account the geometry of the shape. This work was undertaken as homework and as such had no teacher input.

The teacher did not have any idea that this student, two or three weeks in to meeting trigonometry in class for the first time, knew about and was interested in aspects of the A level mathematics course. The opportunity to direct his own work gave this student the chance to make his understanding known and, through the website game, 'teach' others.

Chosen problem: Firing a projectile at a pyramid.

The way in which he undertook the task was to

- select these equations
  - Time = 2 u sinα / g
  - Range = u^2 sin2α / g
- create an Excel spreadsheet with these headings: velocity, angle, time, time to greatest height, greatest height, range
- create an Excel spreadsheet which worked out the measurements of the pyramid so that the projectile may be aimed at the centre
- create a link to an online projectile game produced by the student for his classmates.

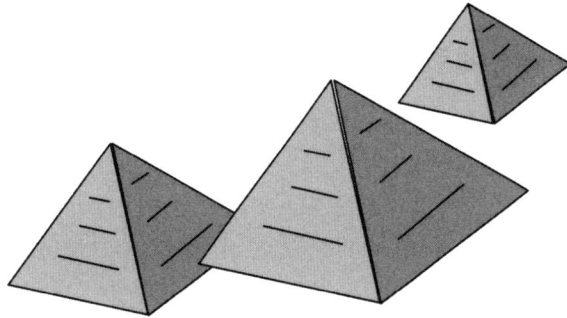

When first completed the Great pyramid was 145.75 metres high. The sloping angles of its sides is 51 degrees and 51 minutes. The pyranid is square in cross-section with its sides lining up with north, south east and west.

The structure consists of approximately 2 million blocks of stone, each weighing more than 2 tons.

The Egyptians also built step pyramids which look like they sound. The height of the Djosier step pyramid was about 62.18 metres with a square base of 125.27 m × 109.12 m.

Explore a pyramid mathematically.

You may work individually or in groups and present your work in whatever way you wish.

CHARLIE'S GARDEN

T | **Teaching Notes**

Prior knowledge

Pupils need to be able
- to apply the mathematics they know in familiar and unfamiliar contexts (exceptional performance)

Lesson objective
- It is not possible to be specific in this instance. Lesson objectives would be set on an individual basis

Suggested outcomes

No suggestions are made to the students but these are a selection of previous responses:

- Three-dimensional models

- Plans and elevations

- Calculations on honey production

- Calculations on water needed

- Spreadsheet looking at birth and death rates for the livestock in the pond

Groupings | Student free choice

Timing | Minimum 1 hour 30 minutes

Equipment
- Internet access or alternative research facilities

- Paper, glue and so on for model-making

Ideally you would first read the story to the student: they may later benefit from being provided with a hard copy. There may be a sense of apprehension initially but as they look at the story in more detail ideas will emerge. You may find that you will need to offer more support than in previous examples.

The story is written so that it offers a wide range of possible responses from plans and elevations, to the use of spreadsheets to look at population growth and decay, to consideration of the normal distribution.

The range is deliberate so that the activity may be used with any of the age group. What it also offers is the opportunity for those gifted mathematicians who do not normally demonstrate their ability to select and develop ideas which motivate them to a high level.

In my experience many students went for a 'safe' option and developed scale drawings, models and plans and elevations. It has to be remembered that as no numbers were given to the students even the simplest idea required considerable thought. Tessellations, bees and honey were also popular themes.

Charlie's garden

It was a warm June day. Charlie walked out into the sunshine and viewed his garden with pleasure. The plot was rectangular, with a pond and several flower beds, designed to create a restful vista from every angle.

The pond was home to both fish and frogs, and Charlie speculated on the number of tadpoles there had to be in spring to keep the stock of frogs constant. His two cats lay in wait in the hope of an extra meal.

Charlie was an old man with children and grandchildren. The garden held many memories for him as he had lived in the same house since he was a child. There were trees he had planted in his youth, then merely twigs but which now towered above him. He wondered about the number and sizes of the leaves. Did they grow to a plan or was it all purely random?

The bees buzzed as they flew from flower to flower in search of the finest nectar. He marvelled at the intricate honeycombs created by the workers. Honey was Charlie's favourite spread and it looked as though he would be able to collect ample supplies this year.

He looked up at the sky. Clouds were gathering. 'No need to water the lawn tonight.'

What mathematics could you use Charlie's garden to illustrate? You may choose your own topic(s), work together with others or on your own and display your understanding in whichever way you prefer.

Out of interest I also read the story to two 14-year-old students who attended A level mathematics classes. They immediately came up with a wealth of ideas – everything I had thought of as I wrote the passage!

This is a clear example of where the teacher cannot be all-knowing and has to work as a facilitator.

Chapter 8

Master classes

This chapter offers a couple of suggestions for master-class activities.

Tasks are:

■ For sale: large horse

■ Mathematics olympics

Master classes may be run in a range of formats such as a single year group across schools or across year groups within one school. They may be run within or outside the school day but, importantly, there needs to be quality time to devote to any activity with the appropriate support mechanisms in place. The numbers need to be big enough to generate a buzz but still be manageable.

Selection of the students is often of concern to the teachers I have worked with. There are several approaches. You can choose to

■ select those who are already achieving
■ select those who are obviously underachieving
■ select a mixture of achievers and underachievers
■ invite anyone who is interested enough to come along.

No method is perfect and inevitably some able mathematicians will miss the opportunity.

If you are able to run such activities on a fairly regular basis then you could choose to use a different process each time and so cast your net wider.

There may be Local Education Authority consultants in your area who would be willing to help with the arrangements and to be an extra pair of hands on the day. If there is a Gifted and Talented Co-ordinator in your school, you could expect help from that teacher and he/she will probably have links with experts in the community.

It can be very productive if for example Year 9 (13 to 14-year-olds) students act as facilitators for Year 5 (9 to 10-year-olds) as they are able to introduce pupils to secondary school mathematics at the same time as developing their own leadership and organizational skills. As the secondary school students undertake the planning it is likely that their mathematical understanding will also be enhanced.

In contrast to the school day, no bells should ring to move students from one area of interest to another, so genuine in-depth interaction with the task can be fostered. Working with the complexity is more important than completion.

Any of the earlier activities could be used as master-class activities but for the two specific examples following there is greater need for a continuous period of time. It would be extremely difficult, though not impossible, for students to be working individually on the task as a whole, although they may naturally break out of their groups for particular tasks.

FOR SALE: LARGE HORSE

T | Teaching Notes

Prior knowledge

Pupils need to be able to

- calculate lengths, areas and volumes in plane shapes and right prisms (Level 7)

- enlarge shapes by a fractional scale factor, and appreciate the similarity of the resulting shapes (Level 7)

- calculate lengths of circular arcs and areas of sectors, and calculate the surface area of cylinders and volumes of cones and spheres (exceptional performance)

Lesson objective

To encourage pupils to

- apply the mathematics they know in familiar and unfamiliar contexts (exceptional performance)

Suggested outcome

- An advert (in any form) to sell the horse

Groupings

Small groups of 3 or 4 with similar strengths

Timing

Minimum 4 hours

Equipment

- IT facilities both for research activities and presentation

- Paper, glue, scissors

There is a considerable range of mathematical knowledge and skills needed to complete this task. Solids used might include spheres, cylinders, cones, and so on, all of which will have to fit together to create a horse. For younger participants you may find it necessary to supply some basic formulae so that they can use a wider range of shapes. If you have wooden solids available this may also be helpful.

Groups may find it more effective to allocate tasks to individuals or pairs within the groups: your knowledge of the students will be most useful in facilitating this process.

Some of the shapes will serve only a decorative purpose but the main body must be sufficiently large, when scaled up, to hide soldiers the size and number of which the students must decide. While the design process is itself quite complex, it is the decisions regarding the scaling up which need to be made clear as there is no definitive number of men to be hidden in the body of the horse. Students decide how and in what numbers the men are to be packed in.

Any advert must make the accommodation clear!

The means by which the horse is moved might also feature as a discussion issue, as might building costs and so on.

At the end of the session, groups can present their adverts to a panel of potential 'buyers'. The panel can comprise a mixture of teachers and students who have not been directly involved.

The buyers have to make a decision and give mathematical reasons for their choice.

For sale: large horse

Design a 'Trojan horse' using only mathematical shapes in its construction.

You may, for instance, construct a scale model of the horse from paper or, if you prefer, produce detailed plans for a construction. You may have other ideas of how to show your horse's qualities.

Produce an advert for the *Mythological Media* extolling the properties of your horse to prospective purchases who might be considering a similar attack on their enemies.

MATHEMATICS OLYMPICS

T Teaching Notes

Prior knowledge
Pupils need to be able to

- apply the mathematics they know in familiar and unfamiliar contexts (exceptional performance). They need a wide range of mathematical skills

Lesson objective
To encourage pupils to

- consolidate and deepen their skills

Suggested outcome

- Scheduled television coverage for one day.

Groupings
Small groups of 3 or 4 with similar strengths

Timing
Minimum 4 hours

Equipment

- Students need to be provided with a range of materials including equipment to build three-dimensional shapes and stopwatches to time practice heats.

- Some races may be computerized

There needs to be an initial whole-group session when the students begin to consider the possibilities for events. After this stage you will need to facilitate the division into smaller groups each working on a chosen event.

Small-group work makes this task manageable but the whole group also needs to work together to obtain timings for the day. I suggest that there is a team which co-ordinates the work of the smaller groups, to avoid repetition.

The students will need to trial events on each other and it is likely that students will discover various strengths and weaknesses in performance in regard to particular races. They could consider training programmes that might be needed (peer assessment for learning). Alternatively they may decide to create expert teams!

The final product itself could, of course, be used as master-class material.

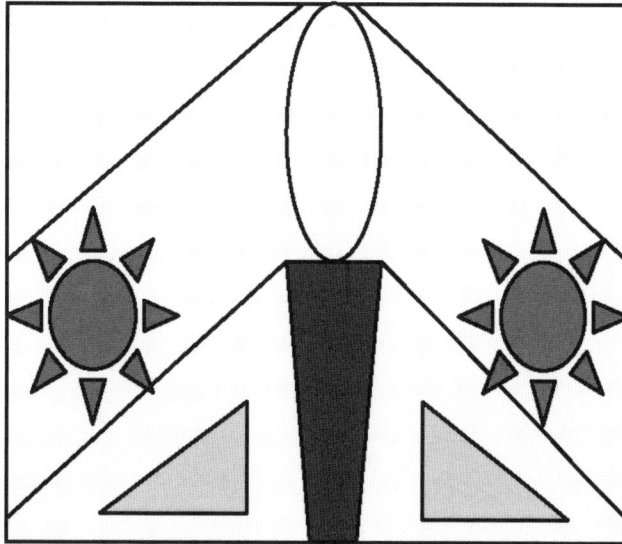

It has been decided that there should be games held every four years to allow mathematicians from across the world to compete against each other in mathematical races.

The races need to reflect the different disciplines in mathematics, so number, algebra, shape, space, and so on all need to be represented. There may be relays and medleys.

Design a series of events. You will need to have prepared the 'races' and have approximations for the times. Unlike track events, heats will have to have equivalent but not identical tasks.

Javelin and shotput could be thought of as open-ended events: can you find mathematical equivalents?

Schedule the television coverage for one day.

Just for Teachers

Are you someone who reads the first few pages and the last page of a book? Please have a look at the middle!

At first glance you may find the tasks rather strange, but if you let your mathematical mind wander off the straight and narrow of the average textbook you will find many exciting detours.

Can I encourage you to think about the activities as you travel to school in the morning or even to pose a version of it as an after-dinner subject for discussion by your family and friends. It's maths, but not as we know it!

If your experience of using this type of material is similar to mine you will soon be caught up in the magic of watching and supporting young minds less restrained by conventional thinking than our own.

Next time you go to a party, explain how exciting it is to be a mathematics teacher and how privileged we are to watch young minds create.

References

Angus, L. (1993) 'The sociology of school effectiveness', *British Journal of the sociology of Education*, 14 (3): 333–45.

Assessment Reform Group (1999) *Assessment for Learning: Beyond the Black Box*. University of Cambridge School of Education. Available online at www.qca.org.uk/294.html

Black, P., Harrison, C., Lee, C., Marshall, B. Wiliam, D. (2002) *Working Inside the Black Box: Assessment for Learning in the Classroom*. London: NFER-Nelson.

Bloom, B.S. et al (1984) *Taxonomy* of *Educational Objectives* © 1984 Pearson Education. Boston, MA: Allyn and Bacon.

Buescher, T.M. and Highman, S. (1990) 'Helping Adolescents Adjust to Giftedness', ERIC EC Digest E489 U.S Department of Education http://ericec.org/digests/e489

Craft, A., Jeffrey, B. and Leibling, M. (2001) *Creativity in Education*, London: Continuum.

Day, C. (2001) 'Innovative Teachers: Promoting Lifelong Learning for All'. In Chapman, J. and Aspin, D. (eds) *International Handbook on Lifelong Learning*, Norwell, MA: Kluwer.

De Bono, E. (1982) *Thinking Course*. London: BBC Books.

De Bono, E. (1985) *Six Thinking Hats*. Boston. London: Little, Brown and Company.

De Bono, E. (1992) *Serious Creativity*. Glasgow: Harper Collins.

Eyre, D., (1997) *Able Children in Ordinary Schools* London: David Fulton.

Fisher, R. and Williams, M. (eds) (2004) *Unlocking Creativity: Teaching Across the Curriculum*. London: David Fulton.

Freire, P. (1996) *Pedagogy of the oppressed*. New York: Continuum.

Freeman, J. (1991) *Gifted Children Growing Up*. London: Cassell.

Galton, F. (1869) *Hereditary Genius: An Inquiry into its Laws and Consequences*. Macmillan: London.

Gardiner, A. (1987) *Discovering Mathematics – The Art of Investigation*, Oxford Science Investigations.

Gardner, H. (1994) *Creating Minds: An Anatomy of Creativity Seen Through the Lives of Freud, Einstein, Picasso, Stravinsky, Eliot, Graham and Gandhi*. New York: Basic Books.

Geake, J. (2004) http://www.brookes.ac.uk/publications/research_forum/vol2_issue1/giftedchildren/initial

Geake, J. G. and Hansen, P. (2005) 'Neural correlates of intelligence as revealed by fMRI of fluid analogies', *NeuroImage*, 26 (2): 555–64.

George, D. (1997) *The Challenge of the Able Child*. London: David Fulton.

Getzels, J. and Jackson, P. (1962) *Creativity and Intelligence*. New York: Wiley.

Guildford, J.P. (1950) 'Creativity'. *American Psychology*. 5: 444–54.

Hopkins, D. and Harris, A. (2000) *Creating the Conditions for Teaching and Learning*. London: David Fulton.

Humphreys M., and Hyland T. (2002) 'Theory, practice and performance in teaching'. *Educational Studies*. 28 (1): 5–15.

Jensen, E. (2000) *Brain Based Learning*. San Diego: The Brain Store.

Joyce, B., Calhoun, E. and Hopkins, D. (2002) *Models of Learning-Tools for Teaching*. Buckingham: Open University Press.

Maker, C.J. and Neilson, A.B. (1995) (2nd ed) *Teaching Models in Education of the Gifted*. Texas: Pro-Ed.

Maslow, A.H. et al., (1999) *Toward a Psychology of Being*. Chichester: John Wiley & Son.

McNiff, J. (1993) *Teaching as Learning: an action research approach*. London: Routledge.

Muijs, D. and Reynolds, D. (2001) *Effective Teaching: evidence and practice*. London: Paul Chapman Publishing.

National Curriculum Handbook (Dfes/QCA). http://www.ncaction.org.uk/creativity

Renzulli, J.S., and Reis, S.M. (1985) *The schoolwide enrichment model: A comprehensive plan for educational excellence*. Mansfield Center, CT: Creative Learning Press.

Slavin, R. E. (1995) *Cooperative Learning: Theory, Research and Practice*, 2nd Edition. Boston: Allyn and Bacon.

Sternberg, R. J.(ed) (1999) *A Handbook of Creativity*. Cambridge: Cambridge University Press.

Sternberg, R. J., and Lubart, T. I. (1991) 'Creating creative minds'. *Phi Delta Kappan*. 72(8), 608–14.

Treffinger, D.(ed.) (2004) *Creativity and Giftedness (Essential Readings in Gifted Education)*. Thousand Oaks, CA: Sage Publications.

Torrance, E. P. and Goff, K. (1990) *Fostering Academic Creativity in Gifted Students*. ERIC EC Digest No. E484.

UK Mathematics Foundation, (2000) *Acceleration or enrichment: serving the needs of the top 10% in school mathematics*. University of Birmingham.

Urban, K. K. (2003) 'Toward a componential model of creativity'. In D. Ambrose, L. M. Cohen, and A. J. Tannenbaum (eds), Creative intelligence: Toward theoretic integration (pp. 81–112). NJ: Hampton Press.

Webb, N. (1991) 'Task-related verbal interaction and mathematics learning in small groups'. *Journal for Research in Mathematics Education*. 22: 366–89.

Index

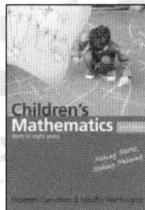

NEW MATHS TITLES FROM PAUL CHAPMAN PUBLISHING

These three books have been developed by the influential team at The Open University's Centre for Mathematics Education for their Graduate Diploma in Mathematics Education.

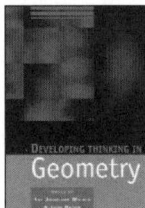

Developing Thinking in Geometry

Edited by Sue Johnston-Wilder and John Mason both at The Open University

'All readers can use this book to reignite their fascination with mathematics' - Professor Sylvia Johnson, Sheffield Hallam University

This book enables teachers and their support staff to experience and teach geometric thinking. As well as discussing key teaching principles, the book and accompanying interactive CD include many activities that encourage readers to extend their own learning, and consequently their teaching practices.

2005 • 288 pages
Hardback (1-4129-1168-0) / Paperback (1-4129-1169-9)

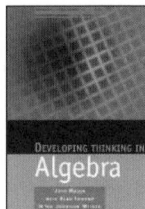

Developing Thinking in Algebra

John Mason , Alan Graham and Sue Johnston-Wilder all at The Open University

Algebra has always been a watershed for pupils learning mathematics. This book will enable you to think about yourself as a learner of algebra in a new way, and thus to teach algebra more successfully, overcoming difficulties and building upon skills that all learners have. This is a 'must have' for all teachers of mathematics, primary or secondary, and their support staff.

2005 • 336 pages
Hardback (1-4129-1170-2) / Paperback (1-4129-1171-0)

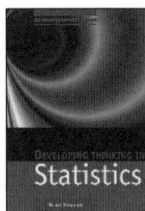

Developing Thinking in Statistics

Alan Graham The Open University

'By getting the reader to think statistically, Alan Graham has provided us with a very readable book that helps to dispel the poor reputation that statistics has acquired over many years' - Professor Neville Davies, Director, Royal Statistical Society Centre for Statistical Education, Nottingham Trent University

Statistics is a key area of the school mathematics curriculum where maths and the real world meet. This book will enable teachers and others interested in statistical thinking to become inspired by the big ideas of statistics and, in turn, teach them enthusiastically to learners.

2006 • 288 pages
Hardback (1-4129-1166-4) / Paperback (1-4129-1167-2)

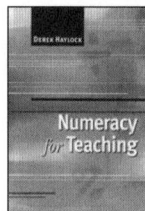

Numeracy for Teaching

Derek Haylock Consultant for the Teacher Training Agency

'The stated aim of this book is to help teacher-trainees prepare for the numeracy test all new entrants to the profession now have to pass. Any trainee worried about the test should find this a useful resource. As in similar books by Derek Haylock , the mathematical content is written in a clear and accessible style' - Mike Askew, Times Educational Supplement

2001 • 268 pages
Hardback (0-7619-7460-1) / Paperback (0-7619-7461-X)

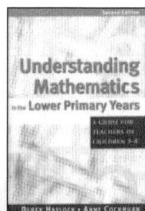

Understanding Mathematics in the Lower Primary Years

A Guide for Teachers of Children 3 - 8

Second Edition

Derek Haylock Consultant for the Teacher Training Agency and Anne D Cockburn University of East Anglia, Norwich

'This book is an excellent example of material designed to help develop our understanding of mathematics and what we, as teachers, can do to better support mathematical understanding in children' - Mary Lynes, Primary Practice, Journal of the National Primary Trust

This is a fully revised and updated edition of the authors' successful and much-used book. The revisions introduced in the Second Edition reflect recent changes such as the Early Learning Goals for the Foundation Stage and the particular emphasis of the National Numeracy Strategy.

2002 • 238 pages
Hardback (0-7619-4103-7) / Paperback (0-7619-4104-5)

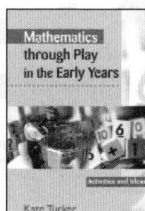

Mathematics Through Play in the Early Years

Activities and Ideas

Kate Tucker Early Years Teacher and Trainer, Exeter

Drawing directly on the classroom practice of the author, this book gives teachers exciting ideas for using play to teach early years mathematics. Written with particular emphasis on transition from the Foundation Stage to Key Stage 1, this book links practical ideas for teaching through play with the Early Learning Goals (ELGs) for the Foundation Stage and the National Numeracy Strategy (NNS).

2005 • 160 pages
Hardback (1-4129-0393-9) / Paperback (1-4129-0394-7)

Visit our website at
www.paulchapmanpublishing.co.uk
to order online and receive free postage and packaging!

Paul Chapman Publishing
A SAGE Publications Company

These books are available to lecturers teaching appropriate courses. To request an inspection copy visit www.paulchapmanpublishing.co.uk/inspectioncopy